The Institute of Biology's
Studies in Biology no. 34

© R. H. M. Langer 1972

First published 1972
by Edward Arnold (Publishers) Limited,
25 Hill Street,
London, W1X 8LL

Boards edition ISBN: 0 7131 2362 1
Paper edition ISBN: 0 7131 2363 x

Printed in Great Britain by
William Clowes & Sons, Limited
London, Beccles and Colchester

General Preface to the Series

It is no longer possible for one textbook to cover the whole field of Biology and to remain sufficiently up to date. At the same time students at school, and indeed those in their first year at universities, must be contemporary in their biological outlook and know where the most important developments are taking place.

The Biological Education Committee, set up jointly by the Royal Society and the Institute of Biology, is sponsoring, therefore, the production of a series of booklets dealing with limited biological topics in which recent progress has been most rapid and important.

A feature of the series is that the booklets indicate as clearly as possible the methods that have been employed in elucidating the problems with which they deal. Wherever appropriate there are suggestions for practical work for the student. To ensure that each booklet is kept up to date, comments and questions about the contents may be sent to the author or the Institute.

1971
Institute of Biology
41 Queen's Gate
London, S.W.7

Preface

Cereals and pasture grasses are economically the most important agricultural plants in the world. Rice, wheat, millet and rye form the basis of the daily diet of many millions of people; maize, barley and other cereals are used widely for livestock production, and a whole host of pasture grasses in many different climates is grazed by animals to provide us with meat, milk, butter, hides and wool. 'All flesh is grass.' No wonder that scientists have studied this essential group of plants intensively. Cereals have been investigated for very many years, but systematic research on pasture grasses is of more recent origin. Much is known, more remains to be discovered. This book attempts to summarize our present knowledge. It deals with the basic growth pattern of the grass plant, how it flowers and produces seed, the ways in which it is affected by environment and adapted to climatic conditions, and how through a better understanding of physiological principles we can improve production from pasture grasses and cereals.

1972
R.H.M.L.

Contents

The Grass Plant 1

Before we can discuss the growth of the grass plant and how it is affected by environmental conditions, we must get orientated, so to speak, by getting to know its component parts. In the vegetative condition, before there is any sign of flowering, the grass plant consists of a collection of shoots, or tillers as they are called (Fig. 1–1). A tiller is made up of a number of foliar organs, each composed of a leaf blade or lamina and a leaf sheath, which arise from nodes at the base of the tiller. The internodes separating the nodes are highly contracted, so that the true stem is extremely short and in fact not visible from the outside. What appears to be a stem to a casual observer is really a collection of laminae and leaf sheaths tightly rolled or folded one inside the other. It is from inside this composite structure that new leaves appear and unfold. In the axil of each leaf, situated at a node, there is an axillary bud which under suitable conditions may grow out to become a new tiller. A further morphological feature worth noting in many grasses is that at the junction of the lamina and the leaf sheath there is the ligule, a membranous structure varying in size and outline depending on the species. In some grasses the base of the lamina is prolonged into two claw-like projections, the auricles.

Flower formation is accompanied by rapid elongation of the upper internodes, and this results in the appearance of a long flowering stem or culm which, in contrast to the vegetative stem, is clearly visible. Attached to its nodes are further leaf sheaths and laminae, and it terminates in the inflorescence.

Basically there are two types of inflorescence, a spike which is unbranched, or a panicle which is branched (Fig. 1–2), although there is also an intermediate type called a spike-like panicle. On the axis of the spike, or rachis as it is called, and on the branches of the panicle there occur groups of flowers, the spikelets (Fig. 1–3). At the base of each of these is normally a pair of bracts or glumes. One or many individual flowers or florets make up a spikelet. Each floret is subtended by another pair of bracts, the lemma (the lower) and the palea (the upper). Florets are joined to one another by a short stem, called the rachilla. The floral parts consist typically of three stamens, two styles terminating in feathery stigmas, and an ovary with a single ovule. At the base of the floret there are two tiny protuberances, the lodicules, which by swelling open up the flower and thus allow the anthers and stigmas to emerge. Despite this device, many grasses are normally self-pollinated, although there are many more in which cross-pollination occurs through wind.

Now that we have acquired a generalized picture of the grass plant and have learnt some necessary vocabulary, we can begin to study its life history from germination and vegetative growth to flowering and seed

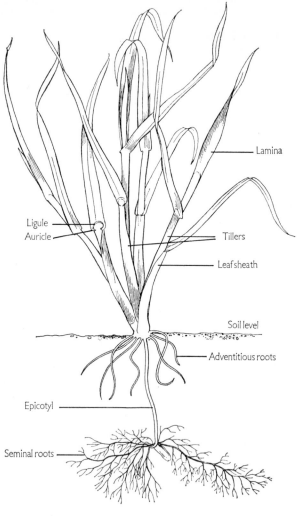

Fig. 1–1 Diagram of a grass plant. (From THOMAS and DAVIES, 1964.)

production. This will involve learning something about the morphology and physiology of grasses and, because they are of such great economic importance, we shall attempt to stress the relevance of botanical features to practical considerations in producing cereal grain, grass seed, or herbage for grazing animals. The connection between science and practice should become even clearer in the final chapters, in which we shall discuss the importance of recent physiological work and the many uses of cereals and herbage grasses.

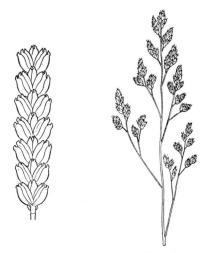

Fig. 1–2 Types of grass inflorescence: spike of wheat (*left*), panicle of cocksfoot (*right*). (After NELSON, 1946.)

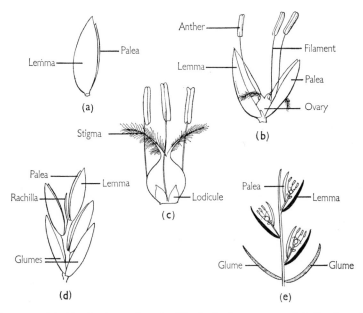

Fig. 1–3 (a) Single floret closed; (b) single floret open; (c) floral parts; (d) spikelet; (e) diagram of spikelet. (From GILL and VEAR, 1969.)

Germination

2.1 Morphological events

In the grasses, what is usually called the seed or grain is really a fruit, known as the caryopsis. Typically, as shown for wheat in Fig. 2–1, it is composed of a comparatively large store of carbohydrates, the endosperm, which is separated from the embryo by a shield-like structure, the scutellum. Just inside the seed coat there is the aleurone layer, a shallow proteinaceous tissue which plays an important part in the biochemical events during germination (see § 2.2) and which influences the quality of cereal grains.

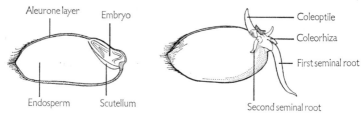

Fig. 2–1 Section through wheat caryopsis (*left*), early stages of germination in wheat (*right*). (From THOMAS and DAVIES, 1964.)

During germination the caryopsis absorbs water and swells. The first outward indication of growth is the appearance of the primary root, the radicle, which bursts through a protective covering, the coleorhiza (Fig. 2–1). Two pairs of lateral rootlets follow in succession. Meanwhile the primary shoot elongates, protected by a colourless sheath, the coleoptile, through which the first leaf pierces and then unfolds. The coleoptile is an organ well-known to all plant physiologists, because it was through its phototropic curvature that auxins were first studied in detail.

The first roots, or seminal roots as they are called, are not the main absorbing system of the grass plant. This function is performed by adventitious roots which arise from the lowermost nodes of the first shoot and the tillers. If the seed has been planted below the surface of the soil, the two root systems are separated from one another through the elongation of one or more internodes which form a kind of rhizome or epicotyl as it is commonly called (Fig. 1–1).

2.2 Biochemical events

The energy for germination and early growth of the seedling must come from metabolites stored in the endosperm, until the leaves have unfolded and begin to photosynthesize. As the seed becomes activated through the absorption of water, there occurs a sharp increase in the rate of respiration as the starch in the endosperm is broken down into simpler carbohydrates. This implies considerable enzyme activity and thus the formation of enzyme proteins. Working with germinating barley seeds at the Waite Institute in Australia, PALEG (1960) has shown that the embryo appears to play a vital part in this process. Enzyme synthesis, especially production of α-amylase, occurs in the cells of the aleurone layer, but in the absence of the embryo these cells do not produce hydrolytic enzymes in response to added moisture. Apparently the embryo provides a hormonal signal which stimulates enzyme synthesis in the aleurone layer, and this has been identified by PALEG (1960) as gibberellin. Concentrations of gibberellins as low as 2×10^{-11} M were found to cause an increase in the content of α-amylase and proteases and activation of β-amylase in seeds from which the embryo had been removed. This response is so sensitive that it is now used as a bioassay for gibberellins. It is interesting to note that in the brewing industry gibberellins are used on a commerical scale to stimulate the germination of barley grain and to ensure the desired conversion of endosperm starch to the simpler carbohydrates of malt.

The seeds of many grass species germinate readily as soon as they are ripe, but there are others which are dormant for varying times and thus require a period of after-ripening. As a general rule we may say that species which are used extensively in cultivation have little or no seed dormancy, but in wild grasses a mechanism of this kind is not uncommon, as it is an important factor in the survival of the species. Even a so-called dry seed contains measureable amounts of water. For example, in wheat this is of the order of 14 per cent. The storage life of seeds is closely related to their moisture content, and care must be taken to prevent it from rising by storing in dry conditions.

3.1 Origin of leaves

Grasses are extremely well adapted to being grazed or cut because, before the flowering stage is reached, leaf formation continues during and after each defoliation. This is because during the vegetative phase the meristematic zones are located close to the soil surface, beyond the reach of animals and machines. Even if some meristems are removed by defoliation, they may readily be replaced by the appearance of new tillers (see § 5.1). Few others plants have such an efficient mechanism of recovery growth, and thus it is no wonder that grasses have assumed such a pre-eminent position as forage plants.

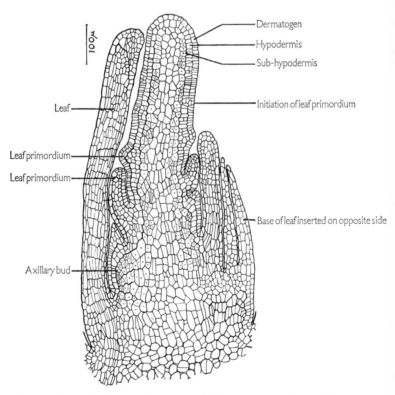

Fig. 3–1 Longitudinal section through the stem apex of couch (*Agropyron repens*) showing leaf primordia at various stages of development. (From SHARMAN, 1945.)

The stem apex (shoot apex, apical meristem, growing point) of each tiller occurs just above the highest node, but since the stem is so highly contracted, its actual position is at the base of the tiller. Careful removal of successive leaves reveals in the very centre the apical dome, an intensely meristematic region. It is along two opposite flanks of this dome that new leaves are laid down in alternating order (Fig. 3–1). Formation of a leaf primordium begins by rapid cell division in the outermost cell layers of the dome, the dermatogen and hypodermis, giving rise to a microscopically visible protuberance. Lateral spread of cell divisions changes the protuberance into a crescent and then a collar, and as this structure grows upwards it assumes the appearance of a cowl when eventually it overlaps the apical dome. Soon after the leaf primordium has encircled the axis, cells in a deeper layer (the sub-hypodermis), just below on the opposite side, start dividing. This activity leads to the formation of a bud which ultimately will be seen in the axil of the next lower leaf primordium which appears on the opposite side from the previous one. The same processes are repeated over and over again as long as the shoot apex remains vegetative, so that at any time it is possible to see a succession of leaf primordia at various stages of development (Plate 1), each of them representing a stage in the sequence of growth through which individual primordia go. The whole region comprising the apical dome and alternating leaf primordia is referred to as the stem apex. The interval between the appearance of successive leaf primordia is called a plastochron, and this is often used as a biological time scale in physiological experiments.

3.2 Morphology of leaf growth

At its inception the whole leaf primordium is meristematic, but soon cell division activity becomes confined to an intercalary meristem at its base. This region becomes divided into two zones through the formation of a band of parenchyma cells, and this coincides with the appearance of the ligule, originally an outgrowth of the epidermis. These events mark the beginning of separate development within the foliar organ, for the upper portion of the meristem is associated with growth of the lamina, while activity in the lower portion leads to growth of the sheath. Cell division and enlargement cause the lamina to move up inside the folded or rolled sheaths of the older leaves. It appears that cell expansion is restricted to the region protected by these sheaths, an environment into which so little light penetrates that the young lamina is unlikely to produce its own assimilates. Emergence of the lamina is accompanied by several profound changes, for not only do the cells of the exposed portion cease expansion but they also encounter an entirely new environment in which they photosynthesize and transpire. Meristematic activity in the lamina comes to an end when the ligule is differentiated, but then the sheath elongates through division and enlargement of its cells, and this continues until the ligule is exposed. This marks the end of elongation growth and the foliar organ has now reached its final length. Of course, the next leaf is meanwhile moving up inside the previous sheath.

We can draw two important conclusions from this course of events. The first is that the tip of a leaf represents its oldest and the base its younges portion. The leaf tip is physiologically more mature and is the first part to senesce when the leaf dies. The second inference reinforces what we have already learnt about the adaptation of grasses to grazing. Not only are leaves initiated on the shoot apex at the base of the tiller, but subsequent leaf growth also originates from meristems which are usually low enough to escape being damaged. Depending on their stage of growth, leaves are affected in different ways by the cutting or grazing of a tiller. The older leaves will have reached their full size, and are thus not able to resume growth. In slightly younger leaves whose lamina has stopped expanding but whose sheath meristem is still active, growth of the sheath will continue, exposing the cut surface. Still younger leaves may have lost only their tip or escaped defoliation altogether, and these will now appear above the level of grazing or cutting. All this can happen very quickly under conditions favouring rapid growth, as close observation of a recently cut lawn will show (Fig. 3–2).

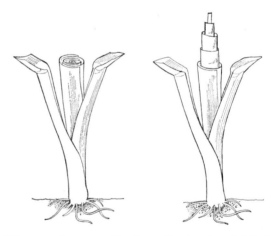

Fig. 3–2   Diagram of a grass tiller immediately after defoliation (*left*) and a few days later (*right*). Note lack of growth by old leaf sheaths on the outside, limited growth by younger sheaths near centre, and at the top appearance of young leaf with tip missing.

3.3   Rates of leaf initiation and appearance

We can conveniently measure leaf growth on a tiller at two times, by determining the rate of initiation as leaf primordia appear on the stem apex, and the rate of appearance as leaves become visible externally and unfold. These two rates are not necessarily the same. In some species and under certain conditions initiation may proceed more quickly than appearance, and consequently leaf primordia accumulate. This is commonly the first sign that flowering is about to begin (see § 6.1) but it can also occur in the absence of floral initiation. SHARMAN (1947) examined comparable tillers of several species and suggested that there may be three groups of species, ranging from those with long stem apices with 15–20 leaf primordia, characteristic of certain pasture grasses, to the cereals which commonly have only a few primordia on a short apex.

Comparatively little research has been done into factors influencing the rate of leaf initiation, presumably because it is merely the prelude to the more obvious leaf appearance. In wheat, raising the temperature from 15 to 20 °C has been found to increase the rate of leaf appearance by over 50 per cent and a plastochron may be as short as 2–3 days. Increasing the light intensity has relatively less effect, but in barley ASPINALL and PALEG (1963) found a linear response of rate of leaf primordium production to light energy between 7.8 and 32.5 W m^{-2} at each of two photoperiods. Also in barley, leaf initiation has been shown to be speeded up by increasing the daylength from 14 to 24 hours at constant light intensity. Information is required on the effect of mineral supply, but water stress has recently been shown to be detrimental.

Our knowledge of the factors controlling the rate of leaf appearance is fairly good though by no means complete and has been summarized in recent reviews by ANSLOW (1966) and SILSBURY (1970). Rates vary within and between species, and there are pronounced seasonal trends. For example, it may take 20 or more days for a new leaf to appear in the winter but only 5 or 6 days in the summer. On the other hand leaves may appear at a constant rate in the more uniform environment of a heated glasshouse. Of the external factors, temperature and light intensity have been thoroughly investigated. Table 1 illustrates rate of leaf appearance in perennial ryegrass (*Lolium perenne*) at different temperatures as reported by several authors.

Many species, wheat for example, show similarly clear responses, but

Table 1   Rates of leaf appearance in perennial ryegrass (number of days per leaf)

Temperature	Other conditions	Rate (*days/leaf*)
25 °C	—	5.8
18 °C	21 530 lux	6.4
12 °C	—	9.4
10 °C	21 530 lux	10.0
winter	unheated glasshouse	15.5
winter	heated glasshouse	9.5

there are others which have a fairly broad optimum temperature range. You will note that even in perennial ryegrass leaves do not appear appreciably faster at 25 °C than at 18 °C, and indeed a range of 18 to 29 °C has been found optimal in other experiments. Several other temperate grass species have been shown to have similar temperature optima, but subtropical species continue to respond to still warmer conditions. Thus paspalum (*Paspalum dilatatum*) produces leaves most rapidly between 29 and 33 °C, but at low temperatures it is inferior to perennial ryegrass. We can use this kind of information to predict the climatic requirements of a species and where it is most likely to grow successfully. Another point which arises from these differences between species is the possibility that the plant breeder may be able to select grasses with rapid rates of leaf production. We shall return to this interesting question later (see § 8.1).

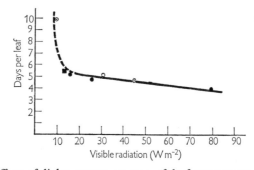

Fig. 3-3 Effect of light energy on rate of leaf appearance in perennial ryegrass (*Lolium perenne*) at 18–23 °C. Results from several experiments. (From SILSBURY, 1970.)

Seasonal fluctuations in the rate of leaf appearance may be caused not only by temperature but also by changes in light intensity and photoperiod. Many experiments with cereals and pasture grasses in controlled environments have shown that leaves appear more rapidly as the light energy increases. SILSBURY (1970) has summarized results obtained by several workers with perennial ryegrass growing at temperatures between 18 and 23 °C (Fig. 3-3). Between 9.7 and 78 W m^{-2} there is a continuous though small decrease in the number of days required for leaves to appear, but at very low light intensity leaf appearance slows down considerably. Although there is not enough information to be sure about the effect of light energy below 9.7 W m^{-2}, we know from experiments with wheat and other grasses that very low light intensity or shading severely limit the rate of leaf appearance.

The effects of photoperiod are less easily evaluated, because in many experiments variations in daylength are confounded with differences in light intensity. To add to our difficulties, photoperiod also affects flowering

in many species, and valid comparisons can only be made while plants are
vegetative. In controlled-environment cabinets RYLE (1964), at Hurley,
found that in cocksfoot (*Dactylis glomerata*) rate of leaf appearance was
slower in a long than in a short photoperiod and, by extending a natural 8
hour day by 8 hours of low-intensity light, leaves of cocksfoot, perennial
ryegrass and meadow fescue (*Festuca pratensis*) also appeared more slowly.
Comparable results have also been recorded in wheat, but there are reports
of opposite effects or lack of response to photoperiod in other experiments
with several species.

Mineral nutrition appears to have little effect on the rate of leaf appearance. Only increasing nitrogen supply has been shown to speed up slightly
the rate at which leaves appear, but there are several other experiments
which have failed to show any nitrogen effect.

Although it is more convenient and informative to examine the effects
of external factors independently in controlled conditions, in a natural
environment these factors are closely interrelated. This can be demonstrated by comparing rates of leaf appearance in different temperatures and
light intensities (Table 2).

Table 2   Rates of leaf appearance in New Zealand ryegrass cv. Manawa.
(After MITCHELL, 1953.)

Temperature °C	10 °		18 °	
Light intensity (lux)	7535	21 530	7535	21 530
No. of leaves per week	0.61	0.87	0.77	1.25

How these two factors interact can be seen by noting that raising the
light intensity had a relatively smaller effect at 10 than at 18 °C. Similarly,
at a light intensity of 7535 lux an increase in temperature speeded up leaf
appearance less than at 21 530 lux, and both factors combined at their
higher level to give the most rapid growth.

One more point needs to be made before we conclude the subject of
leaf appearance, and that is that at any one time only a limited number of
leaves is elongating. The typical pattern in many pasture grasses is that,
while the youngest leaf is just appearing, the one below it is elongating
rapidly and the next older leaf may just be at the end of its extension
growth. However, this basic pattern varies with species and environment.
For example, in rice it seems that a leaf does not elongate until the preceding one is fully expanded and a new leaf primordium has been formed on
the stem apex. Apart from these elongating leaves, a tiller also bears
several fully expanded leaves but their number is also limited because each
leaf lives for only a restricted period, as we shall see later (see § 3.5).

3.4 Leaf size

Leaf size, or more accurately area per leaf, is also influenced by the environment. In order to assess these effects comparisons must be based on leaves occupying similar positions, for on the same tiller there appear to be differences in the size of successive leaves. The first leaf tends to be the smallest and, until inflorescence formation occurs, successive leaves may be larger and their dry weight at the time of emergence greater. On the other hand, relative growth rate (increase in dry weight per unit dry weight) declines from leaf to leaf. However, the sequence of leaf sizes can be modified by external factors. Among these, temperature, light intensity and mineral supply are the most important.

In an experiment with wheat, area per leaf was found to increase with temperature up to 20 °C and then to decline again beyond 25 °C. Leaf length showed a similar response, but width and thickness decreased as the temperature rose (FRIEND, 1966). The optimum temperature for maximum length in other grasses also appears to be higher than for maximum width. This means that, in general, leaves tend to be shorter and wider in cool conditions, but longer and narrower when it is warm. Since length largely determines area, it follows that leaf size will be greater at high than at low temperature. We can verify these effects by comparing winter and summer leaves on the same plant, although other environmental factors are also involved.

Reduced light intensity, except when severely limiting, usually causes grass leaves to become larger, longer but narrower. Leaf thickness also declines, and thus in shaded situations, such as under trees or close to a hedge, grass leaves may be quite large but low in weight. This is well illustrated by an experiment with perennial ryegrass in which a five-fold decrease in light energy at 20 °C caused an increase in leaf size from 15.0 to 24.7 cm^2 but a decline in leaf dry weight from 73.3 to 55.4 mg. In fact the ratio between the area and weight of leaves, the specific leaf area, is a very sensitive measure of incident light energy and of differences between sun and shade leaves. Although the physiological details of this response are not entirely clear, it appears that the greater leaf size at low light intensity goes some way towards compensating for reduced net photosynthetic rate per unit leaf area under these conditions (SILSBURY, 1970).

Increasing the daylength has been shown to cause the leaves of several species to become larger and longer but not necessarily wider. This appears to be a genuine photoperiodic effect, not connected with variation in light intensity, for interrupting a long dark period of 16 hours by weak illumination for one hour also promotes leaf length and area in grasses whose leaves respond to continuous long days. These adjustments to photoperiod are very rapid and appear to involve increases in cell number and cell size in long days. At the same time, rate of leaf appearance is reduced, as we have noted earlier (see § 3.3).

Supply of minerals also influencee the size of individual leaves. This is particularly true of nitrogen, although the effects on other aspects of growth such as tiller and total leaf numbers are so much greater that little precise information on area per leaf has been recorded. Raising the nitrogen supply does not only increase leaf area but may also modify the succession of leaf sizes on a tiller.

3.5 Leaf senescence and death

To conclude our discussion of leaf growth we must now consider leaf senescence and death. Leaves are organs of limited growth and, once they have reached their final size, they remain on the plant for a certain period and then die. In comparison with many dicotyledonous species the longevity of grass leaves is low. From the scanty evidence that is available it appears that in general the rates of leaf appearance and death do not differ greatly, so that the number of living leaves present on a tiller varies within relatively narrow limits. However, environmental conditions do play a part, because they do not necessarily affect both rates in the same way. For example, in a study of seven different grasses growing during the winter in England, the mean number of living leaves on the main tiller varied from 4.5 to 5.8 according to genotype in a heated and from 3.1 to 3.7 in an unheated glasshouse. Increased nitrogen supply slightly raised this number but only at the higher temperature (RYLE, 1964).

Senescence begins at the tip of the leaf, the oldest part, and then spreads downwards. Cell constituents are mobilized and redistributed, so that the leaf loses weight. Further weight losses are caused by decomposition. If nutrients or water are in short supply or if the leaf is deprived of light through shading, one would expect earlier senescence and death than in more favourable conditions. Particularly in a dense pasture community, leaf death is an important, though often neglected, factor which may depress growth rates.

Another important consideration is that after a leaf is fully expanded its rate of apparent photosynthesis begins to fall, at first slowly for the initial third of its remaining life but then more rapidly as it ages and senescence approaches. Following expansion a leaf exports assimilates to younger leaves, to developing tillers and to the roots. However, with age these contributions become less and they cease altogether with advancing senescence. Although there is no evidence to show that dying leaves are a liability to the rest of the plant, we can say that they are not essential and that they certainly do not contribute assimilates for growth elsewhere. Young, emerging leaves, on the other hand, retain all their assimilates and may receive supplies from older leaves. By analogy we can state that young middle age is the prime of life of a leaf, when it is fully grown but before senescence begins.

Stem and Root Growth 4

4.1 Nodes and internodes

During their early development the successive leaf primordia produced by the stem apex are inserted closely above each other without being separated by internodes. Separation occurs later through cell division in the region between adjoining primordia. Initially this involves all cells but, as the internode grows, meristematic activity becomes restricted to the basal portion. A basal intercalary meristem is thus formed. As long as the stem apex remains vegetative and continues to produce leaf primordia, the total amount of growth by internodes is normally very small, so that successive nodes are closely packed. However, cell division and enlargement may bring about internode elongation independently of flowering in those grasses which form stolons or rhizomes. In fact we can say that the creeping habit of grasses depends not only on the orientation of shoots, which are said to be diageotropic, but also on the ability of internodes to elongate without flowering having occurred. Not all internodes participate in the process of elongation. Those at the base of the rhizome remain short and the buds in the axil of their scale leaves may produce shoots. More recently formed internodes may elongate, but there appears to be a curious pattern whereby long and short internodes adjoin one another or one long internode alternates with several very short ones (BARNARD, 1964). Grasses with a tufted growth habit, notably the cereals, have also been reported to show some stem elongation in the absence of flowering. This may occur in short days at very low light intensity and can probably be interpreted as an adaptation to such unfavourable conditions.

The main signal for internode elongation to occur is the advent of the reproductive stage (see § 6.1). In this case also not all internodes are involved and the basal ones remain short. Normally only 4 to 5 upper internodes elongate, and this occurs very rapidly giving rise to a culm or flowering stem. The leaf sheaths attached to the nodes in the same region also grow rapidly at this time. The whole process is referred to as shooting or jointing, especially when cereals and pasture grasses are grown for seed.

The vascular bundles of the grass stem are surrounded by sclerenchyma and resemble those of other monocotyledonous plants in having a closed collateral arrangement. The numerous vascular traces from the lamina pass into the sheath and then further downwards into the node. However, instead of fusing immediately with traces coming from higher up in the stem they curve inwards and pass unbranched through the node. The first connection with other vascular tissue occurs usually at the next lower node. By contrast, the vascular traces of axillary buds are joined at the node of their insertion where a complex nodal plate develops. We thus have the unusual situation that the bud and the tiller arising from it (see § 5.1) have a more direct relation with the leaf at the next higher node than with the leaf in whose axil they arise.

4.2 Roots

As we saw earlier (Fig. 1–1) grasses have two root systems, the seminal and the adventitious or nodal roots. The seminal roots arise from primordia that are present in the embryo and vary from one to seven or eight in number according to species. By weight the seminal roots also appear to be insignificant, amounting to less than 5 per cent of the total root mass in annual grasses or in perennials during the first year, but this apparent inferiority hides the importance of these roots. They are much more highly branched than the adventitious roots and thus exploit a greater soil volume than their small size suggests. By allowing the two root systems

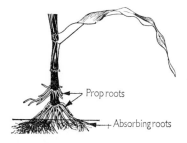

Fig. 4–1   Prop roots arising from the basal nodes of a maize stem. (From WALLACE and BRESSMAN, 1949.)

of timothy plants to take up nutrients separately, WILLIAMS (1962) was able to calculate that the seminal roots absorbed fifty times more nutrients per unit weight than the nodal roots. If all or part of the nodal roots are severed, the seminal roots compensate with additional growth and activity. Because of this observation and the technical difficulties involved in the study of roots, we cannot make any precise statements on the physiological significance of seminal roots during the life of the grass plant. Since they are small and limited in growth, they are unable to support the plant effectively over a long period. In some experiments they have been shown to remain functional throughout the life of annual cereals, although this is not supported by other workers, but in perennials seminal roots are important only during the first months of growth, after which they disappear. As far as our inadequate information goes, it appears that all roots have a limited life span, probably little more than a year at the most. We may thus assume that within the root system there is continuous turnover involving senescence, death, decay and new formation, and that this is connected with the renewal of the tiller population (see § 5.4).

Adventitious roots arise in parenchyma tissue at the nodes just below the internodal intercalary meristem. In tufted grasses this occurs normally at or close to ground level where the internodes are short. However, in some species like maize or sorghum, adventitious roots may appear at nodes well above soil level, serving as supporting rather than absorbing organs, as their name of buttresses or prop roots implies (Fig. 4–1). Stolons and rhizomes also have adventitious roots at their nodes and this enables these stems to ramify as virtually independent units over some distance from their insertion. This in itself is of considerable ecological significance but it becomes even more important, in that rooted portions of stolons and rhizomes are a means of vegetative propagation. This is the basis of the success of couch or twitch (*Agropyron repens*) as a persistent weed and of marram grass (*Ammophila arenaria*) planted as a stabilizer of sand dunes.

It is generally agreed that in temperate species the optimum temperature for root growth tends to be lower than for the growth of shoots. These grasses grow best when soil temperature is relatively low, and under these conditions they can tolerate above optimal air temperatures. In pastures this difference in response has important ecological consequences, for early in spring roots may grow more rapidly than shoots, provided carbohydrate supply is not limiting. This extension of the root system lays the foundation for extremely rapid top growth, once air temperatures begin to rise. On the other hand, hot conditions during the summer may be deleterious through their effects on roots. In subtropical species higher temperatures are required for maximum root growth than in grasses adapted to cooler climates. Roots of maize were found to grow best at 30 °C compared with 20 °C in a temperate species like oat. Another subtropical species, bermuda grass (*Cynodon dactylon*), was shown to be adversely affected by relatively low soil temperature (21 °C) even if the air temperatures were high.

Roots are very sensitive to reduction in light intensity suffered by the leaves. Many experiments are on record which show that reduced light levels or shading have more serious effects on roots than on shoots. This is seen particularly well at low light energy, as in an experiment with smooth brome grass (*Bromus inermis*) in which lowering the light intensity from 30 500 to 1690 lux decreased shoot weight five-fold compared with a thirty-fold decrease in root weight. Light is thus of great ecological significance as a factor in plant competition, because a suppressed species in a dense sward will suffer primarily through having its root growth affected, and this in turn will diminish its ability to take up water and minerals. Carbohydrate supply from the shoots is likely to be the main physiological agency, on the grounds that at low light intensity and high temperature the amount of assimilates reaching the roots is reduced. Decrease in carbohydrate supply is also the main reason why the rate of root growth drops sharply when leaves are removed through cutting or grazing.

Tillering

5.1 Origin of tillers

So far we have been taking a rather restricted view of the grass plant by studying the growth of a single tiller, the unit structure of the plant. Now we must turn our attention to tiller production, because grasses are composed of very many of these shoots. Tillers arise as axillary buds on the stem apex (see § 3.1), as a result of meristematic activity in sub-hypodermal tissue. In the embryo of the seed, buds are usually visible in the axil of the coleoptile and the first one or two foliage leaves. After germination new buds are initiated at the same rate as leaf primordia, usually two or three plastochrons later than the primordium itself. Buds are thus laid down in regular succession from the base upwards, or acropetally as it is called, and they tend to emerge as tillers in the same order. Each bud is the replica of the parent structure, complete with apical meristem, leaf primordia and axillary buds. On emergence the tiller again

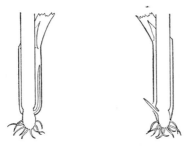

Fig. 5-1 Tiller growing up within leaf sheath in a tufted species (*left*), and breaking through leaf sheath to form a stolon or rhizome (*right*). (From NELSON, 1946.)

resembles the parent shoot with its own system of leaves attached to a highly contracted stem and its own adventitious roots. Although complete in every respect, tillers remain in vascular connection with one another, and there has been much argument as to how independently they function.

A tiller emerges from the encircling leaf sheath in one of two ways, depending on the species (Fig. 5-1). In tufted or tussock-forming grasses the tillers grow upwards within the sheath and first appear externally near the base of the parent lamina. We shall be concerned mainly with this more common type of tillering. Alternatively a tiller may break through the protecting sheath and give rise to a stolon as in rough-stalked meadow grass (*Poa trivialis*) or a rhizome as in couch (*Agropyron repens*). However, at the nodes of their creeping stems these species may produce plants with a tufted growth habit within which the former type of tillering occurs. Whichever the mode of emergence, it depends on the genotype and environmental conditions whether or not a bud develops into a tiller, and this requires to be discussed in some detail (see § 5.3).

5.2 Tiller hierarchies

We have seen that the first shoot, or main tiller, of the grass plant produces leaves, each subtending an axillary bud. If these buds develop further, they will produce in acropetal succession a number of subsidiary shoots which are called primary tillers. These tillers have their own foliage leaves which in turn may produce shoots from their axillary buds, and these are designated secondary tillers. From the leaves of these shoots we get tertiary tillers, and as this process continues a complicated system of tillers of various orders develops on the same plant. A hierarchy of tillers is thus formed (Fig. 5–2), connected by a complex system of anastomosing vascular tissue. One is reminded of the old doggerel:

'Big fleas have little fleas upon their backs to bite them,
Little fleas have lesser fleas, and so *ad infinitum*.'

The analogy with big and little fleas has more meaning than meets the eye, because within the same plant there is considerable variation in the size of tillers. At any one time some will be very small, bearing only a few leaves and possibly no adventitious roots as yet, while others are well established and may have produced several daughter tillers. This gradation is seen particularly well when flowering occurs, in that the main shoot and the first tiller tend to have the largest inflorescence, followed by decreasing size in successive tillers according to their time of origin. It also follows that in general external conditions do not affect all tillers in the same way, because some may be at a more responsive stage of growth than others. This is well illustrated by the observation that very young tillers are often more vulnerable to environmental stress than older ones.

5.3 Factors affecting tillering

Little is known about the growth of tillers before they become visible externally, so we shall confine our discussions to the rate at which tillers appear. Basically, this rate depends on the rate of leaf appearance, although there is a difference in magnitude. Under constant conditions leaves appear at a linear rate, but since this occurs on each tiller and since each leaf has the potential to develop its axillary bud, the rate of tillering tends to be exponential, as long as the plant remains vegetative and the environment favourable.

Before we can assess the effect of external factors, we should note that the amount of tillering is also genetically controlled. Some species produce tillers freely, and indeed pasture grasses are often selected for this charac-

Fig. 5–2 Plant of annual meadow grass (*Poa annua*) showing tillers of different orders, some of them bearing an inflorescence. (From HUBBARD, 1968.)

teristic, while others tiller only sparsely. For example, in an experiment with seven different grasses growing in a uniform environment RYLE (1964) found that at the 10-leaf stage perennial ryegrass had formed an average of 6.2 tillers compared with only 3.3 in timothy (*Phleum pratense*). Timothy seedlings were found to have few if any tillers in the axil of the first three leaves. Other workers have noted that in the same species, five leaves may be present before the first subsidiary tiller appears, whereas in perennial ryegrass a tiller may be seen to emerge as soon as the leaf above it is fully expanded. However, although genotypic differences exist, tillering is highly modified by the environment.

The effect of temperature has been investigated in many grasses. The general conclusion appears to be that in temperate species the optimum temperature is relatively low, ranging from approximately 18–24 °C in ryegrass to 24–29 °C in cocksfoot. In wheat a rise in temperature from 10 to 25 °C was found to favour leaf production more than tillering, but in subtropical species like *Paspalum dilatatum* there is little decline in tillering up to 35 °C. The inhibitory effect of high temperature appears

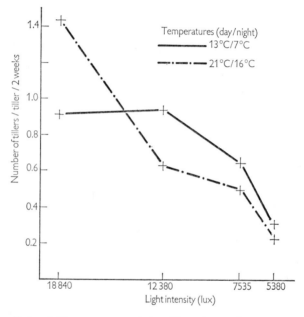

Fig. 5–3 Rate of tiller appearance in seedling plants of ryegrass and prairie grass at two temperatures and in different light intensities (number of tillers per tiller in 2 weeks; both species combined).

to be related to respiration rates and the soluble-carbohydrate content of the plant, since warm conditions at night are often more deleterious than during the day. However, before we begin to discuss possible mechanisms controlling tiller production, we should remember that other environmental factors are also involved and that they interact with one another. This is illustrated by some recent results we obtained with young ryegrass and prairie grass (*Bromus unioloides*) plants growing in controlled conditions. If the light intensity was low, the optimum temperature for tillering was also low. On the other hand, tiller production was most active at high temperatures, provided that light intensity was also increased (Fig. 5–3).

Since these results are based on two grasses with contrasting growth habit, they are given in terms of number of tillers appearing per tiller present at the beginning of a two-week period.

Figure 5–3 shows that the ability of grasses to produce tillers appears to be very sensitive to changes in light energy. A wide range of species has been examined; and without exception high light intensity was found to favour tillering. For instance in an experiment with perennial ryegrass and cocksfoot tiller numbers per plant declined continuously as natural light was reduced from 100% to 5%, and in wheat an increase in light intensity from 7.6 to 95.4 W m^{-2} favoured tillering more than leaf production. Grasses appear to be sensitive to current rather than previous light conditions, for ryegrass plants transferred from shade to full light were found to resume tillering at the same rate as others which had never been shaded. Buds which have been prevented from developing by reduced light energy or other unfavourable conditions are apparently not easily stimulated to resume growth by a return to an optimal environment, and many fail to produce a tiller, if the dormant period has been prolonged.

Because of concomitant effects on light energy and flowering, our knowledge of whether photoperiod influences tillering is not at all clear. In some experiments in which these complications appear to have been avoided there is a suggestion that short days encourage rate of tillering in several grasses, although there are reports to the contrary in at least one species, tall fescue (*Festuca arundinacea*). We are not well informed on the effects of water supply, despite the importance of this factor. However, from experiments with sideoats grama (*Bouteloua curtipendula*) we know that amount and rate of tillering are reduced under dry conditions, and in perennial ryegrass and timothy resumption of tillering after flowering is delayed in the absence of rain or irrigation.

We are on much safer ground when it comes to mineral nutrition. In both cereals and pasture grasses, tiller production is greatly increased by raising the supply of nitrogen, phosphorus and potassium (LANGER, 1966), and limitations of other essential elements would also be expected to have an effect. This is hardly surprising, for tillering implies intense meristematic activity and cell enlargement. Of the major elements, nitrogen seems to be the most important. For example, in an experiment with timothy twice as many tillers appeared within three weeks at 150 ppm of nitrogen as compared with 6 ppm, but it took another four weeks before the same difference occurred in plants maintained at high or low levels of phosphorus. Although potassium caused consistent effects, a doubling of tiller numbers was not achieved at all. Nitrogen was also found to interact significantly with the other two elements, in that at low levels of N the plants did not respond to either P or K. These nutrient effects, which have been found consistently in many experiments, show that tiller production can be increased in the field, although the magnitude of response to fertilizers will depend on the nutrient status of the soil and on the stage of development of the plant. Another important implication of these results is that with an increase in tiller numbers more leaves will be produced and, as we saw earlier (see § 3.4), leaf growth itself is also stimulated. Consequently total leaf area per plant is greatly increased.

A most instructive series of experiments on the involvement of nutrient supply in the control of tillering in barley has been reported by ASPINALL (1961). This concerned not only the number of tillers produced, but also the periodicity of the process, for in both cereals and pasture grasses tillering is normally arrested around the time of rapid stem elongation, after the reproductive phase has begun, but it may be resumed when inflorescences have emerged. When all nutrients were applied at the

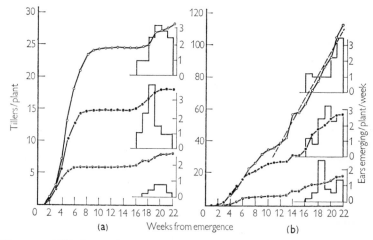

Fig. 5-4   Tillering and ear emergence of barley (cv. Piroline) under varying nutrient regimes. (a) All nutrients applied before germination: ○ 100 per cent; ● 50 per cent; ◖ 10 per cent. (b) Nutrients replaced weekly; solution: ○ 50 per cent; ● 5 per cent; ◖ 1 per cent. Histograms: number of ears per week emerging. (After ASPINALL, 1961.)

beginning of the experiment, before germination had occurred, tillering followed a distinct two-phase pattern (Fig. 5-4). At a low level of nutrition, tillering stopped sooner and after fewer tillers had appeared than in plants which were given nutrients at full strength. Nutrients applied regularly every week had the effect of reducing the low-tillering phase, depending on their concentration. The same nutrient solution diluted to 5% and supplied weekly gave more continuous tillering than when applied at full strength at the beginning of the experiment only and a large dose applied regularly led to virtually uninterrupted tiller production, despite the fact that stem elongation had occurred.

On the face of it, one might be tempted to suggest that tillering is controlled simply by the supply of metabolites, both mineral nutrients, as we have just seen, and assimilates as indicated by the light energy effects described earlier. However, there are many results which cannot be explained on this simple basis. For example, the depressing effect of stem elongation on tillering does not appear to be a function of mineral supply alone, for it occurs in the presence of abundant nutrients which even in annual cereal plants are now known to be taken up well beyond this stage. Only continuous feeding of plants with relatively concentrated solutions, as in Aspinall's experiment, seems to overcome this effect, and there are good grounds for believing that the internal allocation of nutrients is at least as important as their external supply. There is evidence to show that developing grains in the barley inflorescence compete strongly for minerals to the detriment of bud growth and that it takes disproportionately high concentrations of external nutrient supply to counteract this influence. Some other internal regulation of tillering is implied, and this could well take the form of hormonal control either directly or, more probably, through the flow of metabolites. The possible rate of growth regulators, first suggested over twenty years ago, has recently been demonstrated in experiments with wheat and perennial ryegrass in which triiodobenzoic acid (TIBA), known to inhibit auxin transport, has been shown to stimulate tiller bud growth. We have obtained similar effects with kinetin in wheat, while gibberellins usually depress tillering. The situation is not at all clear and it is part of the general problem of apical dominance which remains a riddle, despite many attempts to solve it in dicotolydonous plants which in this respect are more amenable to experimental treatment than grasses.

5.4   Tiller life histories and interdependence

Whether a grass is an annual, biennial or perennial depends very much on the longevity and life cycle of its component tillers. Annual grasses, including most cereals, are the simplest case, for they produce tillers over a restricted period. These tillers either form inflorescences or they die, and since conditions are not normally conducive to renewed tillering after grain setting, these plants have an annual habit. Only if inflorescence emergence is prevented by repeated cutting, and provided nutrients and water are supplied in optimal amounts, can the life of these annuals be prolonged.

At the other end of the scale there are the perennial grasses which produce tillers over prolonged periods, even though there are distinct seasonal trends. Tillers thus appear in a variety of environmental conditions. Some behave like those on the annual grasses, in the sense that they appear, flower and die within the same season, thus describing an annual life cycle. Other tillers on the same plants tend to flower in the year following their formation in the manner of a biennial plant, while others again stay vegetative and remain alive for periods ranging from a few weeks to many months. At a time of environmental stress many young vegetative tillers tend to die soon after they have appeared, but under favourable conditions life may be prolonged to a year or more, as long as flowering does not occur. Work in England with swards of timothy cut at the hay and aftermath stage has shown that tillers arising straight after flowering and during the following weeks form the bulk of the sward for the remainder of that season and carry the plants through the winter. Later formed tillers, although prominent in early spring, are unlikely to survive the summer and most tend to die while still vegetative. The earlier tillers, however, produce flowers in the summer and form the basis of the seed or hay crop. New tillers then arise, and the annual cycle is repeated.

Although other species may not behave in quite such an orderly manner, the general principle emerges that a grass plant itself can be perennial with no definite upper limit to its longevity, but that the individual tiller has a restricted life span. Grasses may thus be perennial only by virtue of their ability to renew their own tiller complement. By the same token the individual plant, and even more so the sward, are in an ever-changing, dynamic condition in which their tiller population is never static. Even if a count indicates the same number of tillers on two consecutive occasions, there may well have been internal change, as some tillers died and others appeared. It is very much like a human or any other biological population in which birth and death occur continually, even though a census may reveal no change in numbers. The size of the population may stay the same, increase or decrease, depending on the rates of death and renewal.

Are tillers nutritionally dependent on one another? This question has been debated for a long time although it is only since the advent of radio-isotope research that much meaningful progress has been made. Earlier studies showed that, as a tiller senesced and died, there was an increase in soluble metabolites following the breakdown of proteins and carbohydrates, and since vascular connections can readily be demonstrated, it was assumed that these soluble substances moved to other parts of the plant.

It is also generally agreed that the requirements for substrate of young buds expanding to form a shoot must be met from elsewhere in the plant, and that the need for minerals continues until the tiller has developed its own adventitious roots, while dependence on external carbohydrate supply stops when the tiller leaves unfold. Work with ^{14}C has confirmed that assimilates move from the parent shoot to young tillers, but between older shoots movement of carbohydrate appears to be more restricted. Based on a thorough investigation of young plants of Italian ryegrass (*Lolium multiflorum*), MARSHALL and SAGAR (1968) concluded that the root system, which received about 70% of the total exported radiocarbon from each shoot, should be considered as the main sink for assimilates. However, there were also considerable export and import of labelled carbon between tillers. The main shoot exported approximately equal amounts to all other tillers, but each tiller was found to contribute more to the main shoot than to its sister tillers. The grass plant thus appears to be a highly organized and integrated system rather than a collection of competing individuals, although tillers compete under conditions of stress.

Flowering

6.1 Morphological changes

In the preceding chapters we have been considering the growth of the grass plant in the vegetative phase. It was important to stress this qualification repeatedly, for the advent of the reproductive phase brings about profound changes in the stem apex, internodes and other parts of the plant. These changes must now engage our attention and, having described them, we must enquire what triggers them off.

The first visible sign of the onset of reproduction is seen in the stem apex which elongates rapidly forming leaf primordia in quick succession. Some idea of the changing rate of development is gained from the observation that in maize plastochron 6 is 4.7 days compared with only 0.5 days for plastochron 13 which closely precedes the formation of floral organs. These new primordia develop only slightly as leaves and, towards the top of the apex, they can barely be seen. Instead, it is the buds in the axil of these primordia which now grow rapidly, giving rise to a double structure composed of leaf and bud primordium. These so-called double ridges are taken as a definite indication that the reproductive process has begun. In physiological terms the most significant event that occurs is the complete change in the relationship of leaf and bud. Whereas in the vegetative condition the bud remains dormant, often for prolonged periods and in many cases never developing into a tiller, the reverse takes place in the reproductive state. The leaves are now inhibited and the buds develop further. It would be most interesting to discover the exact mechanism causing this reversal, for it might tell us something about the suspected hormonal basis of flowering about which we are still largely ignorant.

Once formed the bud primordia develop further at a rapid rate. In grasses whose inflorescence is a spike they become spikelet primordia. In species with panicles they develop into the primary branches of the inflorescence, and this is followed in identical manner by the formation of secondary and subsequent branches, on all of which spikelet primordia appear. The apical dome itself becomes the terminal spikelet in most species with either type of inflorescence and, as soon as this has occurred, there is no further possibility of any more leaf primordia being laid down. Spikelets or primary branches cannot increase in number when this has happened.

Each spikelet primordium differentiates further to form florets with their floral organs (Plate 2). In wheat the glumes and the first and second lemma are seen to arise first as ridges half encircling the axis. With the appearance of the third lemma a floret primordium becomes visible in the axil of the first lemma, and a narrow ridge of tissue on its opposite side heralds the formation of the palea. The floret primordium differentiates further by giving rise to the stamens and the carpel, while the lodicules are the last of the organs to appear. WILLIAMS (1966) has described in detail the developmental physiology of successive floret primordia. In wheat grown at 17 °C and in long days he found that about 8 to 9 days elapsed between the formation of the first and seventh floret. There are slight variations from this basic pattern in other species, depending mainly on the number of florets formed. In some single-flowered species the floret may develop directly from the apical meristem of the spikelet, while in others the apex aborts and the floret is of lateral origin. Whatever the detail, within each floret the stamens continue to grow in size, and the carpel differentiates to form the ovule and finally the two styles.

Consequent upon these changes of the stem apex is the elongation of internodes which we have already noted (see § 4.1). As a result the developing inflorescence is carried upwards within the encircling leaf sheaths, and, soon after the final leaf, the so-called flag leaf, has appeared, inflorescence or ear emergence occurs. The stem apex, originally a minute structure at the base of the plant, has been dramatically transformed into a complex inflorescence elevated high above the leaves. It will be obvious that after floral initiation has occurred on any tiller, the stem apex ceases to produce any further leaf primordia and only those already formed will appear as foliage leaves. Removal of such tillers by cutting or grazing, an easy matter as the stem apex becomes elevated, thus does not result in the loss of the future leaf producing sites. It could in fact be beneficial in encouraging the appearance of new subsidiary tillers and thus enhance leaf production.

6.2 Flower induction

Before flowering can take place most grasses must pass through an obligate vegetative phase. There are a few exceptions, some annual rye-grass species for instance, which may have reproductive primordia on the third or fourth node. Since up to three leaf primordia are present in the embryo, these grasses can flower soon after germination. In other species the vegetative phase may last for a few weeks, as in cocksfoot (*Dactylis glomerata*), meadow fescue (*Festuca pratensis*), and phalaris (*Phalaris tuberosa*), or for several months or even years. Some of the bamboos are examples of the last category. What determines the length of this period is not entirely clear. In some cases it may be the attainment of a certain size or the appearance of certain leaves before flowering can begin, while in others the accumulation of certain metabolites or phytochrome could be important. The length of the vegetative phase can be measured on a time scale or by the number of leaves produced by the stem apex before floral organs are initiated. Even under conditions ideal for flowering a certain number of leaves will emerge, some of them derived from primordia in the embryo and others formed during the period before the onset of reproduction. Although this so-called minimum leaf number may not have the theoretical importance which was once attached to it, we can accept it as a measure of the length of the vegetative phase. The plant is now ready to be induced to flower. The term flower induction has been a subject of some debate among plant physiologists. Some argue that in a strict sense induction has occurred when the plant has reached a stage at which it can respond to photoperiods which will stimulate initiation of reproductive structures. Others include all treatments, including appropriate photoperiods, which ultimately lead to the next phase, flower initiation. To avoid confusion we shall use the term in the latter and wider sense, as it relates to events in the leaf which commit the plant to flower.

The photoperiod is the principal factor which brings about flowering but there are many perennial grasses occurring in the temperate zone which require exposure to low temperature before they respond to appropriate daylengths. Low temperature treatment simulating overwintering under natural conditions is known as vernalization, because it produces the state which the plant attains in the spring. The most effective vernalizing temperatures lie between 0 and 10 °C, well below the optimum for growth. Some grasses, like perennial ryegrass or rye may be vernalized by cold treatment of seeds which have been allowed to imbibe water for a few hours, but in others like *Phalaris tuberosa* the plant must have formed several leaves before it responds. Timothy (*Phleum pratense*) serves as an example that not all temperate perennial grasses have a vernalization requirement. Although the nature of the physiological processes involved is obscure, we know that they occur at the shoot apex itself, because imbibed seeds and even isolated embryos may be vernalized. Once the cold requirement has been fully satisfied, the effects appear to remain stable for a long time even if flower induction does not take place immediately. On the other hand partial vernalization can be reversed by exposure to high temperature, and this is referred to as devernalization.

Vernalization is a quantitative process, since up to a point the longer a sensitive species has been treated, the sooner it will flower. In some genera like *Lolium* the cold requirement varies with increasing perenniality (Fig. 6–1). Although there may be differences among races of the same species, EVANS (1960a) found that perennial ryegrass has an obligate cold requirement and does not flower if vernalized for less than two weeks. Longer treatments give progressively more rapid flowering. Italian ryegrass, a winter annual, flowers without vernalization but responds slightly to cold. The hybrid produced from these two species is intermediate in its reaction, and incidentally it perennates for several years, but the annual *Lolium temulentum* is insensitive to low-temperature vernalization.

Practically all grasses requiring vernalization subsequently need long days for flower initiation (Table 3, p. 31). Not only is this of taxonomic interest, for these species all belong to the sub-family Festucoidae, but it also emphasizes their distribution in the temperate zone where the environment provides these conditions. However, the winters in this region are also characterized by short days, and from the ecological point of view it is therefore no coincidence that in these grasses short days at temperatures

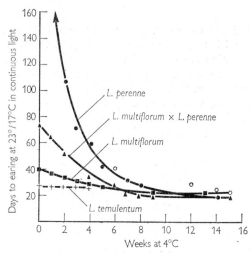

Fig. 6–1 Effect of varying periods of vernalization at 4 °C, given in short days, on the number of days required subsequently for flowering in continuous light. (From EVANS, 1960a.)

normally above 10 °C can replace low-temperature vernalization as a prerequisite for subsequent flowering in long days. Physiologically it is rather difficult to explain why short photoperiods which are perceived by the leaves should have the same effect as exposure to cold which, as we have seen, acts directly on the shoot apex. Be that as it may, grasses like cocksfoot, perennial ryegrass, smooth brome grass (*Bromus inermis*) and browntop (*Agrostis tenuis*) require either vernalization or short days before being able to respond to long days. By contrast, other species in the genus Agrostis react to short days but not to cold, creeping bent (*Agrostis stolonifera*) for example.

   It would be satisfying to be able to say that what we know about flower induction in the grasses has solved the mystery that surrounds the physiology of flowering. Regretfully this is not so, even though some fascinating contributions, notably by EVANS (1969), have come from investigations of grasses, so that gradually the knowledge gap is being closed. We know that it is the leaves which perceive the photoperiodic signal which is then transmitted to the stem apex. It appears that perception of the stimulus must await the attainment of a certain leaf area or the unfolding of a sensitive leaf. Evans favours the second alternative, because in *Lolium temulentum* only 3 to 5 cm² of the sixth or seventh leaf need to be exposed to one long day for full induction to occur, equivalent to the total leaf area of seedlings 14 days old which require 6 long days to be induced. The strain used in these experiments had a critical photoperiod of 9 hours for repeated cycles of long days, but when only one long day was given, 14 to 16 hours were required. The velocity of translocation of the floral stimulus

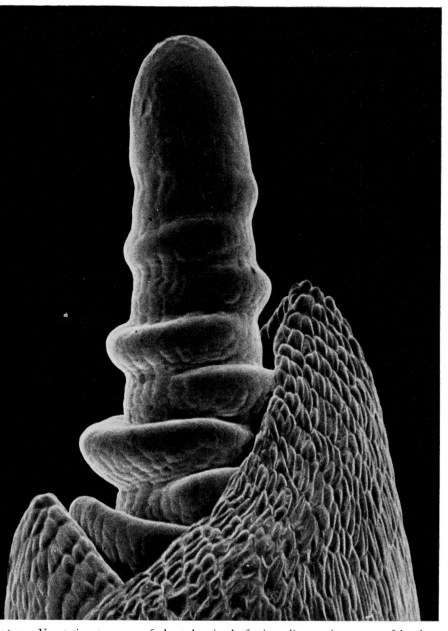

Plate 1 Vegetative stem apex of wheat showing leaf primordia at various stages of development (× 300). (Scanning electron microphotograph courtesy of Dr. J. H. Troughton.)

Plate 2 Reproductive stem apex of wheat showing spikelet and floret primordia at various stages of development (× 40). (Scanning electron microphotograph courtesy of Dr. J. H. Troughton.)

Plate 3 Photograph of a timothy inflorescence showing florets with stigmas and anthers protruding, proliferations with two foliage leaves, and elongated leaf-like lemmas (× 3). (From LANGER and RYLE, 1958. Photograph courtesy of The Grassland Research Institute, Hurley, Berkshire.)

was estimated to be about 2 cm h^{-1}, very much slower than the movement of ^{14}C labelled assimilates which moved at $77–105 \text{ cm h}^{-1}$. Although 2 cm h^{-1} is only an estimate, depending as it does on a subsequent flowering response by the stem apex, the indications are that in this plant the floral stimulus is unlikely to be translocated in mass flow with the assimilates, but it may be significant that the estimated velocity is comparable with that of gibberellin and auxin movement and of protoplasmic streaming. Single applications of gibberellic acid (GA_3) caused flower

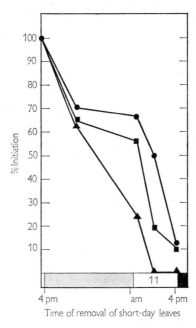

Fig. 6–2 Translocation of the short-day inhibitor. Effect of time of removal of lower leaf blades kept in short-day conditions during exposure of the sixth leaf to one long-day on % plants initiating inflorescences. Long-day leaf area 26.6 cm^2, short-day leaf area 21.1 cm^2 (●), 45.9 cm^2 (■), or 172.4 cm^2 (▲). The hatched area indicates the period when leaf 6 was under illumination while the lower leaves were in darkness. (From EVANS, 1960b.)

initiation in plants held in short days and increased the response to one long-day cycle, but further experiments are required to see whether or not the floral stimulus is in any way related to the gibberellins.

On the other hand Evans has produced evidence indicating the involvement of inhibiting substances in the flowering process. Working with *Lolium temulentum* he found that flower induction was reduced if only some upper leaves were exposed to one long day, while the lower leaves were in short days. The inhibitory effect of short days appeared to be operative outside the leaves, almost certainly at the stem apex itself, because the later the short-day leaves were removed and the greater their area, the more pronounced was the inhibition of flower induction (Fig. 6–2). In other plants in which similar results have been obtained, it has been suggested that reduction in flowering may be a dilution effect, in that the short-day leaves may have diverted assimilates and thus some of the floral stimulus flowing from the induced leaves to the stem apex. However, Evans demonstrated with the aid of ^{14}C that the lower leaves in short days do not act as sinks for assimilates produced by a higher leaf kept in long-day conditions, nor do they reduce the flow of assimilates from this leaf to the stem apex. It thus appears that the leaves exposed to short days produce and export an inhibitor of flower induction. The indentity of this substance is not known, although Abscisin II or some similar compound could possibly be involved. Evans and his colleagues have also studied the biochemical events which occur following induction. Injection of various metabolic inhibitors near the stem apex suggested that ribonucleic acid (RNA) and protein synthesis is essential in induction and that the apex becomes metabolically activated, particularly in positions where future spikelet primordia are situated.

6.3   Daylength and temperature control

The time of flowering in nearly all grasses that have been investigated is determined by the length of day. Only a few exceptions like *Poa annua* are known which are apparently insensitive to photoperiod. Since flowering and seed setting are essential for survival, it is not surprising that species have evolved in regions in which they are photoperiodically adapted. Those which flower only when the photoperiod exceeds a certain critical length are confined to high latitudes. Others which can respond to short days but which flower more rapidly in long days are less narrowly distributed. Grasses occurring in tropical areas tend to be short-day plants and flower only when the days are of less than a certain critical length. Despite the much narrower range of daylengths near the equator, several tropical grasses like some strains of rice can detect changes in photoperiod amounting to less than half an hour per day and thus flower only at a certain time of the year. This geographical response pattern applies not only to different genera and species but also to strains within a species. Thus a Scandinavian strain of perennial ryegrass needs a longer photoperiod for flowering than a strain from the Mediterranean area. The importance of this requirement should be obvious when it comes to the transfer of grasses from one country to another.

Depending then on their distribution, grass species and strains differ tremendously in their response to photoperiod (Table 3). Some have an obligate requirement for long days, others are quantitative long-day plants and flower more rapidly the longer the daylength. Short-day grasses flower when the photoperiod is less than a certain critical length or flower

Table 3 Vernalization and day length requirement for flowering in some pasture grasses and cereals. (After EVANS, 1964.)

o = no response; + = obligate requirement; (+) = quantitative response; (more than one symbol for the same species denotes differences among strains).

	Vernalization requirement	Daylength requirement
Long-day grasses		
Cocksfoot (*Dactylis glomerata*)	+	+
Crested dogstail (*Cynosurus cristatus*)	+	+
Italian ryegrass (*Lolium multiflorum*)	o, (+)	+
Perennial ryegrass (*Lolium perenne*)	(+), +	+
Meadow foxtail (*Alopecurus pratensis*)	(+), +	(+)
Phalaris (*Phalaris tuberosa*)	(+), +	+
Rough-stalked meadow grass (*Poa pratensis*)	+	(+), +
Sheep's fescue (*Festuca ovina*)	+	+
Smooth brome grass (*Bromus inermis*)	(+)	(+), +
Tall fescue (*Festuca arundinacea*)	(+)	(+)
Timothy (*Phleum pratense*)	o	(+), +
Barley (*Hordeum vulgare*)		o, (+), +
Oats (*Avena sativa*)	o, (+)	(+), +
Rye (*Secale cereale*)	o, (+), +	(+), +
Wheat (*Triticum aestivum*)	o, (+)	o, (+), +
Short-day grasses		
Foxtail millet (*Setaria italica*)		(+)
Japanese lawn grass (*Zoysia japonica*)		+
Panicum millet (*Panicum miliaceum*)		+
Sideoats grama (*Bouteloua curtipendula*)	o	o, + ?
Sundan grass (*Sorghum sudanense*)		o, (+)
Switch grass (*Panicum virgatum*)	o	o, + ?
Maize (*Zea mays*)		o, (+), +
Rice (*Oryza sativa*)		o, (+), +
Day-neutral grasses		
Annual meadow grass (*Poa annua*)	o	o
Fingergrass (*Digitaria eriantha*)		o
Needlegrass (*Stipa comata*)		o
Weeping love grass (*Erogrostis curvula*)	o	o

more rapidly in short days. However, the effect of daylength is highly modified by temperature. In many long-day grasses night temperatures above 12–18 °C delay or inhibit flower initiation, especially if the photoperiod is only marginally adequate. For example in an experiment with timothy, floral initiation occurred after 15 days in a 16-hour day at both 13 and 24 °C, but in a 14-hour day it took 79 days at the lower temperature and failed to occur at 24 °C. By contrast, short-day grasses often require night temperatures above 12–16 °C for floral initiation. This difference in response to temperature probably plays an important role in the geographical distribution of grasses, supplementing the effects of photoperiod.

Once initiation has taken place, daylength and temperature continue to play a part in determining the rate of inflorescence development and the rate of stem elongation. In most grasses the time from initiation to ear emergence ranges from 25 to 70 days, depending on photoperiod, temperature, and genotype. Table 4 shows some of these effects in timothy. The intensity of induction has also a pronounced effect, in that the rate of inflorescence development increases with the number of inductive cycles and with the length of the photoperiod to which the plant has been exposed.

Table 4   Estimated number of days between initiation and ear emergence in plants of timothy grown in different photoperiods and temperatures. (From RYLE and LANGER, 1963.)

Temperature (°C)	Photoperiod (h)	
	16	24
13	48	33
18	28	18
24	20	14

This interesting interaction between photoperiod and temperature affecting the date of ear emergence has also been examined by COOPER (1952) who planted clonal propagules of perennial ryegrass in different localities in the British Isles ranging from Scilly in the south to Shetland in the north. The relative differences in time of flowering between early and late strains was maintained at all stations, although there was a general geographical trend from early ear emergence in the south-west to late ear emergence in the north-east. Cooper was able to explain the effects of locality and season in terms of spring temperatures after the critical daylength had been reached. In the early-flowering strain S.24 floral initiation occurred in early March, but the date of ear emergence was shown to depend on the March and April temperatures prevailing at each site. The late-flowering strain S.23 initiated floral structures in early April, but time of ear emergence was determined by April and May temperatures (Fig. 6–3).

Another instructive example of how environmental control of flowering leads to the adaptation of grasses to local climates has been provided by COOPER and MCWILLIAM (1966). Eight Mediterranean populations of *Phalaris tuberosa* collected from a range of latitudes and from habitats varying in winter temperature and distribution of rainfall were planted in Canberra (35 °S) under uniform conditions. Ear emergence dates, recorded over two seasons, ranged from mid-November to mid-December (southern hemisphere). The interesting point which appeared, however, was that

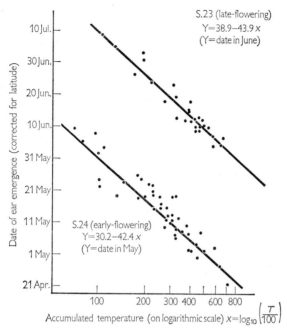

Fig. 6–3 Relationship between spring temperatures and the date of ear emergence in the summer in perennial ryegrass (*Lolium perenne*). March-April temperatures are given for the early-flowering strain S.24, and April-May temperatures for the late-flowering strains S.23, both on a logarithmic scale. (From COOPER, 1952.)

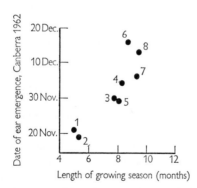

Fig. 6–4 Relationship between date of ear emergence in Mediterranean populations of *Phalaris tuberosa* at Canberra, Australia, and the length of the possible growing season in the original habitat. (From COOPER and MCWILLIAM, 1966). Origin of populations: 1. Israel, 2. Morocco (high altitude), 3. Morocco (low altitude), 4. Algeria, 5. Spain, 6. Greece, 7. Italy, 8. Turkey.

time of ear emergence was directly related to the length of the possible growing season of the populations in their home habitat (Fig. 6–4). Temperature and daylength appeared to be involved as control mechanisms. Plants coming from the warm and dry areas of Morocco and Israel have no vernalization requirement and a relatively short critical photoperiod and are thus able to flower early and take advantage of a short growing season. In the moister and cooler regions of northern Greece and Turkey, potentially with a longer season, plants have evolved in which flower initiation is delayed by a cold requirement and an apparently longer critical daylength until conditions are favourable for seed production.

6.4   Inflorescence development

The environment does not only determine the time of ear emergence but also the size of the individual inflorescence. For any particular genotype the number of spikelets and florets varies according to external conditions, particularly those prevailing soon after floral initiation.

As we have seen earlier (see § 6.1), the first visible sign of floral initiation is the appearance on the stem apex of double ridges from which spikelets develop, if the inflorescence is a spike, or primary branches, if it is a panicle. The number of these structures, which in either case determines the number of spikelets, will depend on how many double-ridge positions

Table 5   Effect of photoperiod and temperature on the number of spikelets and florets per spikelet in perennial ryegrass. (After RYLE, 1965.)

Photoperiod (h)	12.5		20	
Temperature (°C)	13	23	13	23
No. spikelets per ear	27.3	23.5	18.5	17.8
No. florets per spikelet	8.8	8.1	7.8	4.0

are available and thus ultimately on the size of the vegetative stem apex. This is seen most clearly in perennial grasses grown for seed by comparing tillers of different ages. Tillers arising in the autumn and early winter tend to accumulate leaf primordia and, by the time the days are long enough in the spring for induction to occur, they possess many sites at which spikelets can be formed. Later formed tillers are induced at the same time, but because of their later start, will not have as many sites available for conversion. The largest number of spikelets is thus commonly found in the oldest tillers and the smallest in those which do not arise until the advent of flower-inducing photoperiods. Experimentally it is possible to increase spikelet number by delaying initiation. This can be done by imposing either less than optimal photoperiods or unfavourable temperatures, as shown in Table 5 for perennial ryegrass which for rapid initiation requires long days and moderate temperatures. Part of the reason for this effect appears to be that in conditions marginal for flowering the activity of the apical meristem is prolonged so that additional primordia are formed, from which spikelets develop. Nitrogen has also been shown to increase spikelet numbers, but the effect is not very great and in any case it occurs only at an early stage before the final number of available sites has been determined. Similarly, moisture stress at the time of initiation may reduce spikelet numbers.

In considering the number of florets we shall confine our attention here to the total formed because, as we shall see later, not by any means all of them become fertilized and form a seed (see § 7.2). How many florets are produced on each ear will in the first place depend on the number of spikelets, but not all spikelets are equal in this respect. Those near the lower part of the spike tend to contain more florets than those near the top, and the same applies to the branches of the panicle, in either case more especially in the older tillers. Moreover, floret numbers per spikelet can be affected by photoperiod and temperature, as shown for perennial ryegrass in Table 5. Raising the nitrogen supply may also result in more florets being formed without necessarily increasing the number of spikelets. However, the major effect of these factors is on the proportion of florets becoming fertile rather than their total number. Quite apart from these environmental effects the number of florets per spikelet varies from species to species. For example, wheat has typically 8 to 9, sweet vernal grass (*Anthoxanthum odoratum*) has 2, one fertile, the other sterile, and perennial ryegrass up to 14.

Before concluding our discussion of environmental effects on inflorescence development, we should turn our attention to vegetative proliferation. The inflorescence of timothy illustrated in Plate 3 was obtained by exposing plants to only a short period of long days, adequate for floral initiation, followed by a return to short-day conditions. Floral structures were laid down normally, although ear emergence was greatly delayed. However, after the inflorescence had appeared, the plants reverted to partial vegetative growth by producing leaf-like lemmas and, at least in some florets, complete vegetative shoots which were found to be capable of rooting when brought into contact with moist soil. At the same time all floral structures were present, except that no seed was set. Such vegetative proliferations are often seen in normally seed-bearing grasses during moist and warm autumns, and it would appear that under natural conditions late tillers may experience enough long days to start the reproductive process, without being able to complete it as the days become progressively shorter. This alternative method of propagation can be interpreted as yet another environmental adaptation. Proliferations have also been noted, although less frequently, in short-day grasses developing in lengthening photoperiods. There are, however, certain species growing in arctic or alpine habitats, *Festuca vivipara* and *Poa alpina vivipara* for example, in which vegetative proliferations are the normal form of propagation. Although this ensures propagation in an unfavourable environment there is the disadvantage, which also applies to all apomictic grasses, that there is no chance of genetic recombination.

Seed Production 7

7.1 Fertilization and seed formation

Within the grasses there is great variety of breeding systems ranging from apomixis to complete sexuality. Many of the annual species tend to be self-fertilized, while many perennials are cross-fertilized and may remain sterile if selfed. However, there are quite a number of exceptions to this general pattern, depending on genotype and environment. An interesting example of environmental modification is the case of the self-fertile prairie grass (*Bromus unioloides*) which opens its florets in short days and with high soil moisture thus allowing the possibility of cross-pollination, while in long days and dry conditions cleistogamous flowering occurs, pollination and fertilization taking place within closed florets. In monoecious species in which, as in maize, male and female inflorescences occur separately on one plant, the development of the terminal male tassel may be greatly reduced in short photoperiods. Some grass genera are dioecious and have separate male and female plants, while others may be gynodioecious with either female or hermaphrodite individuals.

Anthesis, or opening of the floret, is the first outward sign that pollination is about to begin. In most grasses this occurs some days after the ear has emerged, although there are some in which it takes place just before then. The swelling of the lodicules (Fig. 1–3) is the most common mechanism causing the lemma to part from the palea allowing the feathery stigmas to spread out and the anther filaments to elongate. The anthers then dehisce and pollen is liberated either within the floret or partially or entirely outside it, depending on the type of fertilization. Pollen dispersal from a floret is restricted to several brief periods each day, lasting some 20 minutes in wheat and often slightly less in other grasses. Although pollen is released throughout the day, it occurs most frequently during certain hours. These times seem to vary with the species, but appear to be independent of atmospheric humidity, even though movement of the floral organs is probably caused by changes in the water content of the air. Some grasses shed pollen most profusely in the morning, others at midday, in the evening or at night, and this is repeated for a number of days. Although grass pollen can travel long distances through the air, as sufferers from hay fever will know, it does not remain viable for long in bright sunshine.

Because of its economic importance, grain formation has been studied intensively in wheat. Fusion of the male gametes with the ovum occurs some 30–40 hours after pollination. After fertilization the ovum shows little sign of change for some hours, but the primary endosperm nucleus begins rapid division immediately. During the next few days there is continuous and quick multiplication of endosperm nuclei which initially remain scattered within the embryo sac. Eight to ten days after pollination a continuous layer of cells appears around the embryo sac, and just before this the deposition of starch begins. When the grain has reached about one half of its final size and cell division has ceased, the outermost endosperm cells form the aleurone layer. Meanwhile the fertilized ovum, after a rest of a few hours, also starts dividing, forming initially two cells, the suspensor and an upper cell from which the embryo develops. Further divisions of this cell in all directions now occur in rapid succession, and soon the coleoptile, the plumule, radicle and other structures become visible (Fig. 2–1). Approximately 3 weeks after fertilization the embryo of wheat is fully formed and about 1 mm in length.

Soluble sugars arriving at the endosperm are converted to starch and deposited in granules within specialized plastids inside the endosperm cells. Two sizes of starch granule can be distinguished in the wheat plastid, larger ones formed early in the development of the endosperm cell, and smaller ones which do not increase beyond a certain size. The large granules make up about 90 per cent of the starch by weight but only 10 per cent by number. There are two kinds of starch molecules: amylose, a straight-chain polymer of glucose units, and amylopectin with branched chains of similar units. These molecules are laid in a radial direction in the granule and are joined by hydrogen bonds.

Environmental conditions, during the various stages of grain development appear to be important in determining the ultimate size of the grain. In a recent study with wheat WARDLAW (1970) counted over 64 500 cells in the endosperm of plants maintained at 37 675 lux of light for the first 7 days after anthesis, compared with only about 54 500 in 6458 lux, both counts being made after 21 days. The rate of cell division, but not the final cell number, was reduced by low and increased by high temperatures during the first 5 days following anthesis. However, high temperatures during starch deposition reduced the weight of grains at maturity. As the grain ripens, it continues to gain in weight, but water content declines rapidly until at maturity it is only about 12–15 per cent.

7.2   Factors affecting the number of seeds

The number of seeds produced by an inflorescence will, in the first place, depend on the number of spikelets formed, and the factors important in this respect have already been discussed (see § 6.4). However, analysis of individual spikelets at maturity, particularly in species with many florets, shows that by no means all floret sites are occupied by grains. The percentage of fertile florets has been estimated in many species and can be very variable, ranging from as low as 25 per cent to more than 90 per cent in field-grown cocksfoot and timothy. One has to be careful not to accept such estimates without question, because they can be based either on the number of florets surviving till maturity or the number originally present at the time of ear emergence. These two figures are unlikely to be the same, because many florets, especially those near the tip of the spikelet, tend to degenerate and are no longer visible by the time seed is mature in the fertile ones. This can be demonstrated clearly in wheat, in which 8–10 florets are formed in each spikelet, although typically only 2–4 grains mature at the base of the spikelet. The infertile florets reach progressively less advanced stages of development, the more distal they are on the spikelet axis, most of them failing to be fertilized and aborting. At the time of harvest hardly any trace remains, except for the lemma and palea of one or two of them. The position of the spikelet also plays a part, because in wheat those at the very vase of the spike are often devoid of grains, and the terminal spikelet is also less fertile. A similar gradation in fertility has been observed in perennial ryegrass and other species.

What controls the proportion of florets capable of setting seed? Part of the answer lies in the genetic make-up of the plant, because different cultivars of wheat tend to have different numbers of grains per spikelet under comparable conditions, but the many processes occurring between anthesis and grain ripening are also sensitive to environmental influences. This applies particularly to stress conditions, such as extremes of temperature or reduced water supply. For example, LANGER and OLUGBEMI (1970) found that exposure of wheat plants to 40.5 °C for only 3 hours during anthesis reduced mean grain number per spikelet from 2.3 to 0.4, and to 0.8 in plants given frost treatment at minus 2 °C for the same period of time. Both male and female reproductive structures appear to be sensitive to temperature injury, and from work with barley and Russian wild rye (*Elymus junceus*) it would appear that frost may injure and kill pollen grains, while styles may shrivel under the influence of heat. However, this apparent distinction may disappear in extreme conditions. Water stress at critical stages of development has also been shown to cause increased floret sterility in several species. The involvement of mineral supply has been investigated by LIEW (1968) who in wheat studied the effects of high (150 ppm) and low (15 ppm) levels of nitrogen between the double-ridge stage and the appearance of spikelet primordia and between spikelet formation and ear emergence. It is clear from Fig. 7–1 that high nitrogen supply throughout both periods gave greatly increased floret fertility, particularly in the most basal spikelets which at low N contained little or no grain. If high N was maintained for only part of the time, it appeared to be more beneficial early rather than later in development when spikelet primordia had appeared and differentiated into florets. Variation in nitrogen supply after ear emergence had no additional effect in any of the treatments. From Fig. 7–1 we can also see that, irrespective of nitrogen level, the most basal and terminal spikelets contain fewer fertile florets than those near the middle of the spike.

The suggestion has been made that failure of some florets to form grain may be due to competition for a limited supply of assimilates. That this is not likely to be the main explanation has been shown by RAWSON and EVANS (1970) working with wheat. In the variety Triple Dirk no more than two grains occurred in the central spikelets of plants grown in high light

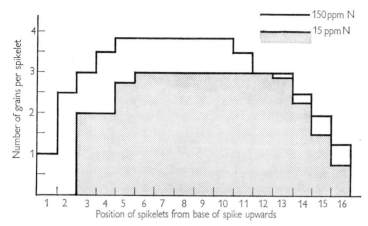

Fig. 7–1 Effect of high (150 ppm) and low (15 ppm) levels of nitrogen supply between the double-ridge stage and ear emergence on the number of grains formed in successive spikelets on the wheat spike. (After LIEW, 1968.)

intensity, although up to four florets reached anthesis. If the lowest floret was sterilized at ear emergence, compensation by the third floret forming grain took place, and sterilization of both the first and second floret resulted in grain formation in nearly all fourth florets. However, competition for available assimilates was probably minimal at the time, and so it appeared that some interaction between florets, probably hormonal in nature, was involved. A similar influence may have been responsible for preventing the fifth and later florets from reaching anthesis, but here we must await further research before this important question is fully understood.

A further factor affecting the number of seeds which has to be considered is seed shedding, losses occurring from ripe inflorescences. This is a serious problem in grasses, such as cocksfoot, ryegrass, timothy and many others. Losses of 50 per cent or more have been reported, depending on the weather before and during the harvest. Part of the problem stems from the fact that within the same inflorescence seeds vary in maturity, and there is additional variation between inflorescences and different plants. Earlier harvesting than customary goes some way to reducing losses, but there are distinct possibilities of selection by plant breeders for improved seed retention. This has been shown in *Phalaris tuberosa* in which the seed lies free at maturity, but certain genotypes have been found whose glumes remain unopened thus preventing the seed from being shed. Another example is certain timothy selections in which the single-flowered spikelets are more firmly attached than in others.

So far we have been considering factors affecting the number of seeds within individual spikelets or on the same inflorescence. However, a very important topic still to be discussed is the number of ear-bearing or fertile tillers, a subject of great concern especially in perennial herbage grasses. The grass plant, as we have seen, consists of a collection of tillers forming an integrated system with a certain amount of nutritional interdependence (see § 5.4). Certainly when it comes to ability to flower and the size of the inflorescence, the position of the tiller appears to be very important. This was shown clearly in an experiment with timothy in which primary tillers were found to have a much higher chance of producing an ear than secondary tillers arising at the same time and in the same environmental conditions. These plants, grown from seed, were in their first year, but similar positional effects seem to play a part in field crops of perennial species, because labelling of tillers has shown that, in general, those arising in early autumn, soon after the seed harvest, form the bulk of the ear-bearing tillers at the next harvest. For example, in cocksfoot over 70 per cent of the inflorescences may be contributed by tillers which were present in the previous autumn, as is shown in Table 6 for a range of nitrogen and density treatments. The physiological reason for the superiority of early tillers is not at all clear, although it has been suggested that it may be related to their greater size, measured by larger leaf weight and the presence of roots.

Table 6   Percentage composition of a crop of cocksfoot ears at harvest in terms of the time of origin of contributing tillers. (After LAMBERT, 1963.)

Month of tiller origin (Northern Hemisphere)	% Contribution to crop of ears
August or earlier	73.3
September	3.4
October	1.8
November and December	12.0
January	3.0
After January	2.9

Certain environmental conditions appear to affect the ability of tillers to form an inflorescence, even if all requirements for low-temperature vernalization and appropriate photoperiod have been satisfied. Mineral nutrition, for example, seems to play an important part. In studies with single plants of several species it has been found that severe lack of nitrogen will prevent ear production, although vegetative growth may continue at a low rate. Increasing the nitrogen supply stimulates tiller and inflorescence production. The percentage of fertile tillers does not necessarily increase at the same time, but this is hardly surprising because many small tillers in unfavourable positions are also produced. Moreover, greater vegetative growth may lead to internal competition for other environmental factors which may prevent greater response to nitrogen. This could well be part of the reason why in the field nitrogen does not always increase and sometimes decreases the proportion of tillers forming ears. One of the factors for which competition ensues appears to be light energy. This has been demonstrated by RYLE (1967) who found that reducing the light intensity greatly reduced the fertility of cocksfoot tillers, caused a smaller reduction in meadow fescue, and had only a slight effect in perennial ryegrass (Table 7).

Table 7   Effect of light intensity on the fertility of the main stem in perennial ryegrass, meadow fescue, and cocksfoot. (After RYLE, 1967.) (Full light = 100.)

Relative light intensities	% Fertility of main stem		
	Perennial ryegrass	Meadow fescue	Cocksfoot
100	100	100	67
50	100	80	40
25	100	79	13
5–10	88	0	0

Ryle has suggested that this effect of light intensity may explain, at least in part, why low population density is known to be desirable for optimum seed production in several grasses. In the field several techniques have been employed with success, such as a low seed rate, row culture and the creation of gaps in the row by mechanical means. These measures may well improve the light climate of the tillers and, in addition, reduce competition for water and minerals.

7.3   Sources of carbohydrate

Few other topics in the physiology of the grass plant have created greater interest and stimulated more research in recent times than the source of the carbohydrates deposited in the grain. When it became apparent that the numerical analysis of grain yield components, particularly in cereals, provided no real explanation of the physiological basis of yield, attention was turned to the fundamental processes involved in the development of grains. The rate of photosynthesis, the relative importance of assimilating organs, and the movement of assimilates became the focal point of many investigations.

Research earlier this century has shown that cereal grains continued to accumulate nitrogen until one week before harvest. Although mineral absorption by the plant may continue until maturity, even if the rate declines after ear emergence, a high proportion of grain nitrogen is obtained by translocation from the leaves, stem and, at a later stage, from the glumes and other parts of the inflorescence. It was, therefore, not unnatural to assume that carbohydrate accumulation by the grain might follow a similar pattern. However, it was soon discovered that the stems and roots of barley did not lose much dry weight and sugars while the grain was filling, and increasingly evidence came to light indicating that photosynthesis after ear emergence accounts for most of the carbohydrate in the grain, with only minor contributions coming from stored sugars in the stem. Since the number of functional green leaves decreases rapidly as the cereal plant approaches maturity, it also became clear that the leaves alone were unlikely to provide enough assimilates and that other green parts of the plant were also involved. In order to estimate the relative importance of the contributing organs—the uppermost leaves and sheaths, the upper

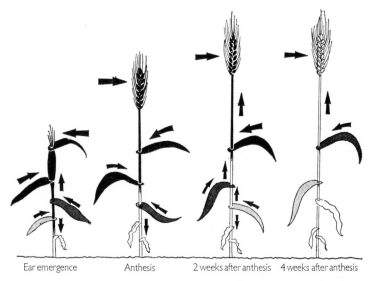

Ear emergence   Anthesis   2 weeks after anthesis   4 weeks after anthesis

Fig. 7–2 Photosynthesis and distribution of assimilates in a maturing cereal plant. Depth of colour indicates increasing rate of photosynthesis, size of arrows indicates magnitude of assimilate transport. (After STOY, 1966.)

portion of the stem, and the ear itself—many experiments were conducted in which one or more of these were prevented from photosynthesizing by shading or defoliation. Subsequent work, however, showed that estimates obtained in this way were not reliable, for photosynthesis in the remaining organs tended to compensate for losses elsewhere, and it was not until radiocarbon and direct measurements of photosynthesis were employed that unequivocal evidence became available (Fig. 7–2).

Exposure of wheat ears and flag leaves to $^{14}CO_2$ followed by autoradiography or quantitative determination of radioactivity has shown beyond any reasonable doubt that the ear itself photosynthesizes actively but tends to retain its own assimilates and that the carbon assimilated by the flag leaf is translocated predominantly to the developing ear (Fig. 7–2).

THORNE (1965) working at Rothamsted in England has constructed a balance sheet accounting for gains by photosynthesis and translocation as well as losses by respiration, using the formula

$$W = P_E - R_d - R_n + S$$

where W represents the final grain weight, P_E carbohydrate contributed by ear photosynthesis, R_d loss by respiration of the ear by day and R_n by night, and S carbohydrate derived from assimilation in the shoot including the flag leaf, sheath and upper part of the stem. Based on the assumption that all CO_2 absorbed is converted to hexose, P_E in wheat was estimated as between 17 and 30 per cent, while R_n amounted to 8–11 per cent and R_d probably ranged from 10 to 33 per cent, suggesting that respiration of the ear exceeds its own assimilation. Shoot photosynthesis, S, estimated as about 115 per cent, appeared to compensate for these losses. Expressed in terms of total carbohydrate entering the wheat ear, 17 per cent appeared to come from ear photosynthesis and 83 per cent was translocated from the shoot system. Part of this carbohydrate is lost through respiration, and the remainder provides the starch of the grain. On the same basis, 60 per cent of the ear carbohydrate in barley can be attributed to ear photosynthesis, the remaining 40 per cent being translocated from the shoot. The barley ear seems to photosynthesize actively, as shown by P_E estimates of 67–91 per cent, without, however, incurring greater respiratory losses than wheat. Some average estimates of the sources of final grain weight in wheat and barley are shown in Table 8. In rice most of the grain carbohydrate is also produced after ear emergence and only up to a third may be derived from stored carbohydrate in the shoot. As in wheat, the inflorescence photosynthesizes but the rate is not high enough to compensate for increasing respiration as the grains develop. The flag leaf system is thus the major contributor to grain carbohydrate. In maize, too, grain formation depends on photosynthesis after ear emergence, but the ear contributes very little of this and the sheaths also appear to be relatively unimportant. Nearly all the grain carbohydrate is derived from assimilation in the leaves which, in maize, remain green and functional much longer than in wheat and barley.

Table 8 Estimation of the percentage of the final grain weight supplied from various sources. (After THORNE, 1965.)

	Barley	Wheat
P_E	79	24
R_d	−24	−28
R_n	−10	−11
$P_E - R_d$	(55)	(−4)
$P_E - R_d - R_n$	(45)	(−15)
S	55	115
W	100	100

On the basis of this information, we can now pose the question: what plant attributes are likely to improve grain yield? Clearly, leaf area after ear emergence is likely to be a very important factor. In annual cereals, for instance wheat or barley, leaf area index (leaf area per unit area of soil surface) reaches a peak soon after the ears have emerged and then declines. However, different cereal varieties do not always show the same pattern of leaf area index (Fig. 7–3). Leaf area duration after ear emergence, D, that is leaf area integrated over the period between ear emergence and maturity, is often highly correlated with yield. This relationship can be further improved by including only the area above the flag leaf node. It is also important that the leaves must be well illuminated and suffer no restriction in water and mineral supply.

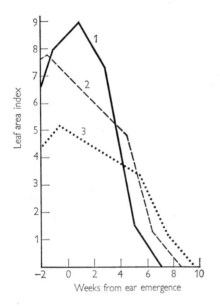

Fig. 7-3   Changes of leaf area index with time in varieties of wheat: 1 and 2, winter wheats; 3, spring wheat. (After THORNE, 1966.)

One might, therefore, be tempted to conclude that a large assimilating surface and the provision of conditions conducive to active photosynthesis should automatically lead to high cereal yields. However, recent information suggests that, in most situations, the plant possesses more than enough photosynthetic capacity. For example, in wheat there may be downward translocation of carbohydrates to the roots, especially from the lower leaves and, if the flag leaf is shaded or removed, the next leaf and the inflorescence are capable of greater photosynthetic rates. It follows that the rate of movement of carbohydrates also depends on the 'sink' size, its actual capacity and the efficiency of biochemical conversion processes. In practical terms, we must thus add a high potential grain number and an efficient enzyme system to the list of requirements for a high-yielding cereal plant.

Physiological Adaptation   8
Explored and Exploited

8.1   Ecological adaptation in grasses

The Gramineae with an estimated 600 genera and over 7500 species are geographically more widely distributed than most, if not all, other plant families. Grasses occur in all continents and in all climates which support the growth of higher plants. Vast areas of the world are covered by prairies, steppes, pampas, savannahs and cultivated grasslands. Despite this profusion of species and ubiquitous occurrence, only about 40 species of pasture grasses and about a dozen cereals have been brought into wide-scale cultivation, based on their actual and potential value for forage or grain production. However, this apparently small number hides an enormous amount of genetical variation, for through a variety of agencies, largely the work of man, these species have become dispersed well beyond the limits of their natural distribution and, through selection processes, they have become adapted to a wide range of climates. Much research has been done to explore this climatic adaptation, partly for its intrinsic scientific interest, but partly also in an endeavour to find superior genotypes for use in particular environments or as breeding material. From isolated beginnings, this work has in recent times grown into scientific plant exploration and introduction on a global scale. As a result, several species that are not cultivated in their natural habitat have assumed considerable economic importance in regions far away from their home. For example, *Phalaris tuberosa*, considered of little value in its Mediterranean home, has become an important pasture species in southern Australia.

Physiological variation within the same species has been studied in many grasses, among them tall fescue (*Festuca arundinacea*). This species is grown successfully in cool-temperate regions, such as the British Isles, and also in the Mediterranean climate of North Africa. The climate of the temperate zone is characterized by fairly warm summers with reasonably high levels of radiation and normally adequate rainfall, while in winter both temperature and light intensity are limiting plant growth. By contrast, solar radiation levels in the Mediterranean are higher, especially in winter, and so are the temperatures, but rain falls predominantly in the winter and summer droughts are pronounced. As the result, Mediterranean ecotypes are active in the winter but tend to be dormant in the summer, while temperate types grow in the summer but are winter-dormant. This difference is well illustrated in Fig. 8–1a which shows leaf growth in a British and a North African tall fescue growing in an unheated glasshouse in England. During the short days of the British winter with its low radiation and cool temperatures the Mediterranean selections increased their leaf area by 60%, whereas the British tall fescue (S.170) showed little change. However, with the advent of longer days, higher light energies and temperatures, these differences tended to disappear. Controlled temperature studies with this species, and also with perennial ryegrass and cocksfoot, have shown that about 20 °C appeared to be optimal for leaf expansion in all selections, but that at 5 °C only the Mediterranean varieties expanded leaves quite actively, whereas northern and continental types became semi-dormant.

Unfortunately, cool season activity is associated with susceptibility to frost. This is shown in Fig. 8-1b by the much greater mortality of North African than British tall fescue exposed to freezing temperatures

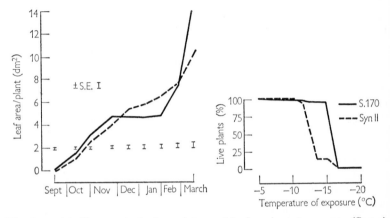

Fig. 8–1 (a) Changes in leaf area/plant with time in a temperate (S.170) and a Mediterranean (Syn II) strain of tall fescue (*Festuca arundinacea*) growing in an unheated glasshouse during the winter; (b) percentage of surviving plants following a 6 hour exposure to a range of low temperatures. (After ROBSON and JEWISS, 1968.)

for several hours. Despite this association, material collected in the Mediterranean area has been used successfully in breeding programmes in the temperate zone, with the intention of improving cool-season production.

Similar possibilities of improving bred varieties of grasses exist when it comes to other characteristics. We have already noted that different requirements for flower induction occur among ecotypes of a single species (see § 6.3), which the plant breeder could use to modify flowering dates and seasonal productivity of existing varieties. COOPER (1966) has reviewed these and other uses of physiological adaptations in plant breeding, including within-species variation in nutrient uptake. A good example is provided by the work of SNAYDON and BRADSHAW (1961) in England who obtained populations of sheep's fescue (*Festuca ovina*) from acid and calcareous soils and grew them in a range of calcium concentrations. Plants derived from acid soils were found to grow better at low Ca concentration than others taken from calcareous soils, thus showing distinct differences in the ability to absorb nutrients from a limited supply. At the other end of the scale there is evidence that *Agrostis* populations growing in the vicintiy of mines are tolerant of copper, nickel, zinc and lead at concentrations which are toxic to plants of the same species not adapted to these conditions. These physiological differences are important not only in relation to ecological adaptation and successful competition with other plants, but they also raise the possibility of plant breeding for improved mineral content and more efficient fertilizer response. It has, in fact, been shown that there is genetic variation involving mineral composition. BUTLER and GLENDAY (1962) studied the iodine content of perennial ryegrass plants obtained from diallele crosses involving parents with high and low leaf iodine. Despite considerable variability, total iodine content was found to be strongly inherited, an encouraging result in view of the importance of this element in animal nutrition (Fig. 8–2).

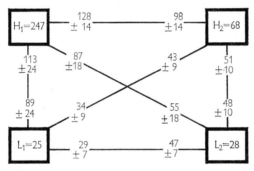

Fig. 8–2   Mean iodine content (g/100 g dry weight), with standard errors, of perennial ryegrass (*Lolium perenne*) plants obtained from diallele crosses involving parent plants with high (H) and low (L) iodine content. Cross means are written adjacent to the female parent. (From BUTLER and GLENDAY, 1962.)

8.2   Variations in photosynthetic activity

The rate of photosynthesis of grasses, in common with all other green plants, depends on the level of light energy. From low intensities of about 500–1000 lux photosynthetic rate rises rapidly in proportion to light energy received, but with higher intensities the rate of increase falls off, until light saturation is reached with no further improvement in the rate of photosynthesis. The saturation light intensity for individual leaves of many temperate species has been found to be around 30 000 lux (Fig. 8–3), although there is considerable variation among genotypes and shade-loving species tend to have lower values. This corresponds to a maximal net photosynthetic rate of 20–30 mg $CO_2\,dm^{-2}\,h^{-1}$. However, within recent years HESKETH (1963) and others have drawn attention to the fact that in subtropical and tropical species, such as maize, sorghum, and *Paspalum*, or Rhodes grass (*Chloris gayana*), light saturation of leaves is not reached till much higher light intensities exceeding 6×10^4 or even 10×10^4 lux. Maximal assimilation rates are also higher ranging from 50–70 mg CO_2 dm^{-2} h^{-1}. Response to temperature also differs between climatic groups, for around 35 °C appears to be optimal for net photosynthesis in tropical grasses compared with about 20 °C in those of the temperate zone. This means that the higher maximum rates in the former are attained only when conditions are suitably warm, and that temperate species usually photosynthesize more rapidly at 20 °C or less.

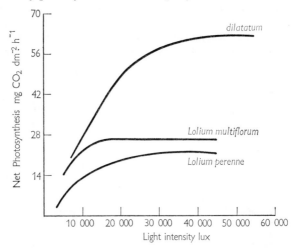

Fig. 8–3 Rate of net photosynthesis in different light intensities in tropical (*Paspalum dilatatum*) and temperate (*Lolium perenne* and *L. multiflorum*) grasses. (From COOPER and TAINTON, 1968.)

The physiological and biochemical characteristics of these two groups of plants have been studied in great detail to account for these differences. It appears that the leaves of some tropical grasses have a comparatively low resistance to CO_2 diffusion, for it seems from an examination of several species that the higher the potential net photosynthesis, the lower is the diffusion resistance of the mesophyll. These tropical species have very low CO_2 compensation points, which means that they can reduce atmospheric CO_2 concentrations to less than 5 ppm, compared with 50–60 ppm in temperate grasses. A further physiological difference is that respiration in perennial ryegrass, cocksfoot and other temperate grasses is stimulated by light, resulting in the release of CO_2, most of which is not recycled in photosynthesis. By contrast, many tropical grasses lack photorespiration and are thus capable of higher rates of photosynthesis. The rapid reassimilation of respiratory CO_2 in tropical grasses is associated with the activity of the enzyme phosphopyruvate carboxylase, resulting in the formation of a 4-carbon acid as the first product of photosynthesis, as opposed to 3-carbon compounds of the Calvin cycle in temperate plants. HATCH *et al.* (1967), the discoverers of this pathway, have demonstrated that per unit weight of chlorophyll this enzyme is four times more active than the ribulose diphosphate carboxylase of temperate species.

Although there are exceptions to this close relationship between climatic distribution and physiological characterization—rice and bamboo, for example, show high photorespiration and lack the 4-carbon pathway— the results of this research mark a major advance in plant science. It might perhaps be possible to enhance the carboxylation capacity of temperate grasses by chemical means, either temporarily or permanently, and thus improve net photosynthesis and yield. A more immediate objective is to use both types of plants to best advantage in intermediate latitudes where temperatures are neither too low for tropical nor too high for temperate grasses. Pastures composed of species with complementary growing seasons and growth potential might provide optimal use of environmental resources. The productivity and energy conversion of temperate and tropical grasses has recently been reviewed by COOPER (1970) who suggests that dry-matter yields of 40 metric tons per hectare are feasible in the tropics, more than twice as much as in the temperate zone, even though low digestibility and low mineral and protein contents may limit its value to animals.

8.3 Mexican wheats and maize mutants

From very many examples that might be chosen to illustrate the extent and use of genetic variation in cereals, we shall briefly consider two recent developments of major importance, one concerning wheat, the other maize.

Some years ago, American plant breeders realized the significance of short-strawed wheats which would respond to heavy rates of fertilizer without lodging. A Japanese dwarf wheat, Norin 10, was introduced and used in a breeding programme resulting in the production of the variety Gaines which was reported to have reached a record grain yield of 14 000 kg/ha in a commercial crop. Subsequently a group of cereal breeders at the International Centre for Wheat and Maize Improvement in Mexico selected more short-strawed lines and from them developed high-yielding wheats suitable for Mexican conditions. By growing two generations of hybrids per year, one near sea level in winter, the other at high altitudes in summer, the breeders speeded up their work and selected for adaptability to a wide range of conditions. These new wheats, together with a great increase in the use of irrigation and fertilizers, almost quadrupled Mexican wheat yields in less than 25 years, an astonishing achievement by any standards. These successes attracted the attention of other countries with food and population problems, and soon Mexican short-strawed wheats were tested and used as breeding material in many parts of the world. The most successful varieties were sown over large areas, and average yields and total grain production started to rise dramatically. In 1968, India virtually doubled its wheat yields, an outstanding event for a country often threatened by famine, and there are now real prospects of self-sufficiency in wheat production for several countries in Asia and elsewhere. No wonder that this development has been hailed as a breakthrough of unusual significance.

The history of maize breeding is often quoted as an excellent illustration of how to exploit hybrid vigour. Double-cross hybrids, obtained from four inbred lines and subsequently crossed, have become the main source of seed for intensive maize production. More recently genetic variation has been further utilized in order to improve the chemical composition of the endosperm for specific purposes. Starch from normal maize consists of 27% straight-chain amylose and 73% branched-chain amylopectin, and these two fractions are not easily separated without incurring high costs. Each kind of starch has its own industrial uses. Amylose is needed for the manufacture of adhesives and thickenings in the food industry, while amylopectin is required by the paper industry and for the fabrication of woven fibre-glass. Through the discovery of endosperm mutants it is now possible to produce crops of maize yielding predominantly one starch fraction or at least a more commercially desirable mixture of the two. The *waxy* mutant produces only amylopectin, and the presence of the amylose extender (*ae*) gene has increased amylose content to 80% of total starch. Another recent advance in maize breeding has been the modification of grain proteins through the use of other mutants. Zein, the prolamine fraction of the endosperm, contains only 0.1 g of the amino acid lysine per 100 g of protein, and the tryptophan content is also low. These two amino acids are essential in the nutrition of non-ruminants including man. Fortunately, two mutations were isolated some time ago with a floury endosperm giving the grain a dull and opaque appearance, and more recently it was discovered that the genes concerned, *opaque-2* and *floury-2*, also increased the lysine and tryptophan contents of the grain. Pigs and rats fed with *opaque-2* maize were found to grow more than three times as rapidly as those given normal maize, and even greater differences were obtained when endosperm meal rather than whole grain was used. The superior nutritive value of this mutant has also been established in experiments with humans. We can readily appreciate how the identification and utilization of mutant genes in maize and other plants could help to solve the food problem of a hungry and overcrowded world.

Cereals and Pasture Grasses in Agriculture

9.1 Distribution and major uses of cereals

Cereals are the main source of food of mankind, either directly as part of the daily diet, or indirectly after conversion into meat, milk, eggs and other animal products. About 70% of the harvested land area of the world, over 700 million hectares, are used for cereal growing. Wheat, rice and maize make up over one half of this total area, and between them they account for nearly three-quarters of the world's cereal production. (Table 9.)

Table 9 World cereal area and production. (After FAO Production Year-book, 1969.)

	Area (hectares $\times 10^6$)	Production (metric tons $\times 10^2$)	Major producing regions
Wheat	227	332	U.S.S.R., Europe, North America
Rice	132	284	Far East, Latin America
Maize	106	251	North America, Latin America, Europe
Millet and Sorghum	111	85	Far East, Africa
Barley	75	131	Europe, U.S.S.R.
Oats	32	54	North America, Europe, U.S.S.R.
Rye	22	33	Europe, U.S.S.R.

Wheat (*Triticum aestivum*) is the foremost cereal of the temperate zone, wherever the climate provides moisture and moderate temperatures for early growth, followed by dry and warm conditions for harvesting. Modern bread wheats are hexaploid (2n=42), the result of a long evolutionary history which included diploid and tetraploid forms and a number of natural crossings with wild grasses. Primitive man probably ate whole wheat florets in the form of soup or porridge, but as early as 2500 B.C. the Egyptians learnt how to grind the grain and to ferment the dough with yeast to produce leavened bread. There are two basic classes of wheat, depending on the physical and chemical characterization of the endosperm. Hard wheats which are grown in continental climates with limited rainfall, have a high protein content and are ideally suited for bread making. Soft wheats, which produce high yields in more humid areas, have a starchy endosperm and are used predominantly for biscuits, pastries and other unleavened products. Tetraploid durum wheats are preferred for the manufacture of macaroni and spaghetti.

Rice (*Oryza sativa*) has also been cultivated since the dawn of civilization and forms the staple diet of millions of people in Asia, parts of Africa and South America. Because of its long cultivation in diverse parts of the world, *Oryza sativa* is a complex species with many forms, probably of subspecies status, most of them diploid ($2n = 24$) but some tetraploids or hybrids. Long-grain rice, grown predominantly in the tropical regions of south-east Asia, is known for its dry, granular consistency after cooking, whereas the short-grain rice of Japan becomes starchy and sticky when cooked. The major part of the world's rice crop is grown in flooded paddy fields which, after cultivation, are planted with seedlings which have been raised separately. About four months later the paddies are drained and the crop is harvested. Large areas of rice are also sown direct after the land has been drained, or by aircraft in flooded fields. After harvesting and threshing the seed is hulled and polished to remove the brown outer layers of the grain, but this results in the loss of protein and vitamin B_1 to the detriment of nutritive value. Rice may also be fermented to form an alcoholic beverage, notably the well-known saké of Japan.

Maize (*Zea mays*) has been described as a more valuable gift of the New World to the Old than all the gold discovered by the European conquerors. From carbon dating of pollen found near Mexico City it appears that maize grew there some 80 000 years ago, spreading to South America and the Caribbean Islands and overland to North America. Nowadays the United States produce over one half of the world's crop, but maize also figures prominently in many other countries with a suitably warm climate. It is customary to distinguish three types of field maize which, depending on the characteristics of the starch, are referred to as flint, dent, and flour corns in decreasing order of hardness. Popcorns have endosperms which on heating expand and invert. Sweet corns, selected for high-sugar endosperms, have become an important vegetable. Maize grain is widely used to feed cattle and pigs and it is also an important source of starch, oil, adhesives and syrup. Silage made from rapidly growing crops of maize is a valuable animal fodder.

Sorghum (*Sorghum vulgare*), another complex species, probably originated in Africa, where it continues to be an important crop. It is also widely cultivated in Asian and American countries with a climate too dry for reliable maize production. Sorghum grain is used mainly for farm animals and poultry, but it also serves as human food, replacing rice, or it may be ground into flour to produce unleavened bread. Millet is a composite term including several genera, such as *Setaria*, *Pennisetum*, or *Panicum*. The main millet producing countries are India, Pakistan and several African states, where the grain is fed to poultry or, in the absence of other cereals, used for human consumption.

Barley (*Hordeum vulgare*), another cereal with a long history of cultivation, is adapted to a wide climatic area of the temperate zone, from the U.S.S.R. to Western Europe, the Middle East, North America and some African and Asian countries. Cultivated barley, a diploid ($2n = 14$), includes basically two types, the two-rowed in which only one third of the spikelets have a fertile floret, and the four- and six-rowed types with all florets fertile. Barley grain is a valuable feed for livestock, particularly beef animals, and only a small amount is used for direct human consumption. However, about a third of the world's crop is used for malting. This process is really controlled germination in which starch is broken down to less complex carbohydrates which subsequently are reduced to simple sugars during brewing and fermented by yeast to yield beer. Malt is also used as flavouring in the food industry and, together with barley grain, it is distilled on a large scale to produce whisky, gin and vodka.

Oats (*Avena sativa*), a hexaploid ($2n = 42$), is grown in cool and moist climates. The grain has good feeding value and is thus highly regarded as a human food, mainly as a breakfast cereal, but also for feeding to horses and other livestock. However, the acreage devoted to oats is declining with the decrease in the horse population. Rye (*Secale cereale*), a diploid species ($2n = 14$), is adapted to cool, dry climates and can tolerate frost. It is thus an important cereal in the U.S.S.R., Poland and Germany, where it is turned into dark, leavened bread, either alone or mixed with wheat. Rye biscuits and rye whisky are other commodities for human consumption, but the grain is also widely used for the feeding of livestock, and in addition rye is valued as a forage plant with considerable cool-season activity.

9.2   Distribution and major uses of pasture grasses

As compared with the cereals, it is infinitely more difficult to obtain an overall picture of the distribution and uses of pasture grasses. Not only are we dealing with a far greater number of genera and species, but also with a wider variety of climates and a greater range of application. Even to estimate the area of the world covered by grasslands is fraught with difficulties, for the published figures exclude sown pastures but include areas of extremely low agricultural value in arid and other inhospitable climates. However, some appreciation of the vastness of our grasslands and of the huge population of productive animals which they support may be derived from the statistics of Table 10.

Within the broad geographical distributions shown in the Table, grasslands occur in many climatic regions, although chiefly where there are discernible differences between wet and dry seasons. In the wet and dry tropical climatic zone, where the mean temperature of the coldest month

Table 10   World distribution of permanent grasslands and of cattle, sheep and goats. (After FAO. Production Yearbook, 1969.)

	Area of permanent meadows and pastures (hectares $\times 10^6$)	*Number of*		
		cattle	*sheep* ($\times 10^6$)	*goats*
Europe	91	124	133	13
North America	280	121	23	4
Latin America	502	244	134	45
Near East	187	40	140	55
Far East	109	252	72	99
Africa	818	130	127	103
Oceania	462	30	227	0

is not less than 18 °C, we find the savannah grasslands. These range from the tropical savannahs of South and Central America to the tree savannahs of parts of Africa and northern Australia. The desert and steppe regions, which together make up more than one-quarter of the world's land surface, occur in dry climates with low rainfall, high evaporation rates, low humidity and, typically, considerable wind. In desert areas the vegetation is sparse, and only isolated areas of bunch-type grasses occur. In the steppe climates these grasses predominate, forming an all-grass flora, typically without legumes and herbs, or a mixed association with scrub trees and thorny bushes. Good examples of grass-steppe are the Argentine pampa or the bunch-grass prairie of North America, while shrub-steppes occur in Patagonia, parts of Australia and elsewhere. The most productive grasslands are found in humid mesothermal climates, in which mean monthly temperatures do not fall below minus 3 °C in winter nor become excessively high in summer, and where there are no serious limitations to growth by continuous drought, heat or cold. Large areas of natural or sown pastures, often following the destruction of forests, have developed in these climates, notably in Europe, North America, Australia, New Zealand and parts of Asia. The fourth climatic zone in which grasslands occur is the humid microthermal with severely limiting winter temperatures. The steppes, prairies, tundras and alpine grasslands of this zone are confined to the high latitudes of the northern and to mountain regions of the southern hemisphere. A more detailed account of the distribution of grasslands has been given by MOORE (1964). HARTLEY (1964) has reviewed the global distribution of grass tribes, genera and species.

In agriculture, grasses are used predominantly for the feeding of grazing animals for the production of meat, milk, wool and other animal products. The act of grazing is a complex process far beyond the simple removal of leaves. It involves physiological changes in all parts of the plant, it alters the microclimate of the pasture such as light transmission, moisture relations and temperature, it changes soil environment through the trampling action of the animal and the return of dung and urine, and it affects the botanical composition of the pasture, particularly the proportion of grasses and legumes, depending on the severity of grazing in relation to the stage of growth of each species. One of these many aspects, the nitrogen cycle in grassland soils, is shown in Fig. 9–1. The art of successful pasture management hinges on a thorough knowledge of the many interacting factors involved, and on how best to match the food requirements of the animal with the maintenance of vigorous pasture growth. Different stocking rates and grazing frequencies, the timing and rates of fertilizer applications, use of irrigation, effective utilization with a minimum of loss, choice of species are some of the many variables in

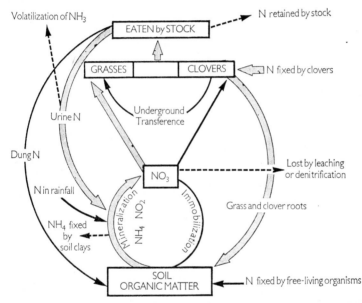

Fig. 9–1 The nitrogen cycle in grazed grass/clover associations, showing the transfer of nitrogen between animals, plants, soil and atmosphere. (From WALKER, 1956.)

grazing systems. Some idea of the magnitude of research in this field can be gauged from the fact that Herbage Abstracts, published by the Commonwealth Agricultural Bureaux, prints well over 2000 abstracts of scientific articles on grassland problems each year. Recent reviews include one by MORLEY and SPEDDING (1968) on agricultural systems and grazing experiments, a detailed account of nutritive value of forages by RAYMOND (1969), and a summary of the effects of grazing on grasslands by MOORE and BIDDISCOMBE (1964). COOPER (1970) has put potential dry matter production from intensively managed temperate pastures at over 20 metric tons/hectare per year, enough for the energy requirements of four dairy cows, while in the tropics twice this dry-matter yield is biologically possible.

In addition to being grazed, grasses may also be cut and conserved for feeding at a later date. The most traditional of these methods is the natural drying of the crop for the production of hay. Unless hay is made rapidly under ideal conditions, a considerable proportion of the dry matter amounting to 24% or more may be lost, and there may also be a serious decline in digestibility and nutritive value compared with the original crop. However, there remains the advantage of a reliable food supply during the winter or other periods of limited grass growth. Another form of conservation is ensilage, during which acids added during the process or produced by bacterial fermentation preserve cut grass which is tightly sealed from the air. There is wide variation in the quality of silage, depending on such factors as the species and its stage of growth, the type of fermentation induced, whether or not the crop was wilted before being ensiled, and the acidity of the silage. In general it appears that the voluntary intake of silage by animals is less than that of the fresh crop and that digestibility may also be lower, but these deficiencies must be viewed against the advantages of having a conserved feed supply available.

Apart from their main function of providing food for animals, grasses are also important in relation to soil conservation and erosion control. Because of their dense and fibrous root system, grasses help to improve soil structure. One of the results of this effect, together with the protection offered by the grass canopy, is the reduced loss of water through run-off and lower soil losses through erosion (Table 11). In arid regions and along sandy coasts erosion by wind is a serious problem. One of the most effective control measures in sand-dune areas is to plant marram grass (*Ammophila arenaria*) whose rhizomes bind and consolidate drifting sand, while its culms and leaves break the force of the wind and cause blown sand to be deposited. Another species worthy of note is the salt tolerant *Spartina townsendii* which plays a useful part in the reclamation of salt marshes.

Table 11   Annual surface run-off and soil losses in relation to type of cover, Missouri, U.S.A. (After COSTIN, 1964.)

	Run-off (%)	Soil loss (metric tons/ha)
Uncultivated soil	48.9	77.5
Wheat	25.2	15.0
Maize	27.4	39.6
Bluegrass sward	11.6	0.7

Finally, we must consider the recreational uses of grasses in lawns, parklands, playing fields and highway landscaping. This is an important and rapidly growing application which calls for specialized knowledge in turf culture and for the production of seed from suitable species. In temperate climates several fine-leaved species of the genus *Agrostis*, *Festuca* and *Poa* are favoured for this purpose, while in warmer regions bermuda grass (*Cynadon dactylon*), bahia grass (*Paspalum notatum*) and carpet grass (*Axonopus affinis*) serve as the main lawn grasses. In some areas, lawns are established from vegetative propagules of bermuda grass, *Zoyzia*, and other species.

9.3   New technology

Modern agriculture has made tremendous progress in raising the productivity of cereals and pasture grasses through the application of scientific knowledge and improved technology. Still greater advances are within our grasp. We have barely begun to exploit some of the exciting possibilities which research has revealed. Scientific plant breeding based on physiological information should bring even greater rewards than past achievements. We are beginning to unravel the complexities of interacting ecological and agronomic factors by constructing mathematical models of various crop systems which we can use to pinpoint deficiencies and to predict crop growth. Along more traditional lines, we can still go further to make the best possible use of fertilizers, irrigation and improved management practices, based on our basic knowledge of how grasses grow.

Another powerful tool for increasing the yield of pasture grasses and cereals is provided by specific chemicals which regulate plant growth. This applies particularly to chemical herbicides whose selective action has been of inestimable benefit to cereal production. Outstanding among them are the phenoxyacetic acids, such as 2,4-D (2,4-dichlorophenoxyacetic acid) and MCPA (4-chloro-2-methylphenoxyacetic acid), which control many annual and perennial broadleaved weeds in cereal crops. As long as reasonable precautions are taken to spray cereal plants at the correct stage of growth, they remain unharmed, while the weeds are checked or killed. By contrast, there are other chemicals which control grasses without damaging other plants, for example dalapon (2,2-dichloropropionic acid). Simazine, a substituted triazine compound, is used specifically as a pre-emergence herbicide for crops of maize, and there are many other herbicides which have become an integral part of crop and pasture management. Some chemicals have in recent years come into prominence because of their modification of plant growth. A good example of this action is the growth retardant CCC, (2-chloroethyl) trimethylammonium chloride, which reduces stem height in wheat and may thus lead to a greater proportion of grain to straw yield (HUMPHRIES, 1968).

In a world threatened by food shortage and famine we have to seek greater agricultural productivity, based inevitably on cereals and pasture grasses as our main source of supply. How grasses grow and how we can influence their growth for the benefit of mankind must be the joint concern of the scientist, the technologist and the farmer.

Further Reading

BARNARD, C. (ed.) (1964). *Grasses and Grasslands*. Macmillan, London.
DAVIES, W. (1961). *The Grass Crop*. Spon, London.
GILL, N. T. and VEAR, K. C. (1969). *Agricultural Botany*. Duckworth, London.
HUBBARD, C. E. (1968). *Grasses*. Penguin Books, Harmondsworth.
LEVY, SIR BRUCE E. (1970). *Grasslands of New Zealand*. Government Printer, Wellington.
MILTHORPE, F. L. and IVINS, J. D. (ed.) (1966). *The Growth of Grasses and Cereals*. Butterworths, London.
MOORE, R. MILTON (ed.) (1970). *Australian Grasslands*. Aust. Nat. Univ. Press, Canberra.
NELSON, A. (1946). *Principles of Agricultural Botany*. Nelson, London.
THOMAS, J. O. and DAVIES, L. J. (1964). *Common British Grasses and Legumes*. Longmans, London.
WHITE, R. O., MOIR, T. R. G. and COOPER, J. P. (1959). *Grasses in Agriculture*. Food and Agriculture Organization of the United Nations.

References

ANSLOW, R. C. (1966). The rate of appearance of leaves on tillers of the Gramineae. *Herb. Abstr.* **36**, 149–55.
ASPINALL, D. (1961). The control of tillering in the barley plant. I. The pattern of tillering and its relation to nutrient supply. *Aust. J. biol. Sci.* **14**, 493–505.
ASPINALL, D. and PALEG, L. G. (1963). Effects of day length and light intensity on growth of barley. I. Growth and development of apex with a fluorescent light source. *Bot. Gaz.* **124**, 429–37.
BUTLER, G. W. and GLENDAY, A. C. (1962). Iodine content of pasture grasses. II. Inheritance of leaf iodine content of perennial ryegrass (*Lolium perenne* L.). *Aust. J. biol. Sci.* **15**, 183–7.
COOPER, J. P. (1952). Studies on growth and development in *Lolium*. III. Influence of season and latitude on ear emergence. *J. Ecol.* **40**, 532–79.
COOPER, J. P. (1966). 'The use of physiological variation in forage-grass breeding', in *The Growth of Grasses and Cereals*, ed. Milthorpe and Ivins. Butterworths, London, 293–307.
COOPER, J. P. (1970). Potential production and energy conversion in temperate and tropical grasses. *Herb. Abstr.* **40**, 1–15.
COOPER, J. P. and MCWILLIAM, J. R. (1966). Climatic variation in forage grasses. II. Germination, flowering and leaf development in Mediterranean populations of *Phalaris tuberosa*. *J. appl. Ecol.* **3**, 191–212.

COOPER, J. P. and TAINTON, N. M. (1968). Light and temperature requirements for the growth of temperate and tropical grasses. *Herb. Abstr.* **38**, 167–76.

COSTIN, A. B. (1964). 'Grasses and grasslands in relation to soil conservation', in *Grasses and Grasslands*, ed. Barnard. Macmillan, London, 236–58.

EVANS, L. T. (1960a). The influence of temperature on flowering in species of *Lolium* and in *Poa pratensis*. *J. agric. Sci., Camb.* **54**, 410–16.

EVANS, L. T. (1960b). Inflorescence initiation in *Lolium temulentum* L. II. Evidence for inhibitory and promotive photoperiodic processes involving transmissible products. *Aust. J. biol. Sci.* **13**, 429–40.

EVANS, L. T. (1964). 'Reproduction', in *Grasses and Grasslands*, ed. Barnard. Macmillan, London, 126–53.

EVANS, L. T. (1969). '*Lolium temulentum* L.', in *The Induction of Flowering*, ed. Evans. Macmillan of Australia, South Melbourne, 328–49.

FRIEND, D. J. C. (1966). 'The effects of light and temperature on the growth of cereals', in *The Growth of Grasses and Cereals*, ed. Milthorpe and Ivins. Butterworths, London, 181–99.

HARTLEY, W. (1964). 'The distribution of grasses', in *Grasses and Grasslands*, ed. Barnard. Macmillan, London, 29–46.

HATCH, M. D., SLACK, C. R. and JOHNSON, H. S. (1967). Further studies on a new pathway of photosynthetic carbon dioxide fixation in sugar cane, and its occurrence in other plant species. *Biochem. J.* **102**, 417–22.

HESKETH, J. D. (1963). Limitations to photosynthesis responsible for differences among species. *Crop Sci.* **3**, 107–11.

HUMPHRIES, E. C. (1968). The beneficial effect of CCC on wheat yields in dry conditions. *Euphytica* Suppl. **1**, 275–9.

LAMBERT, D. A. (1963). The influence of density and nitrogen in seed production stands of S.37 cocksfoot (*Dactylis glomerata* L.). *J. agri. Sci., Camb.* **61**, 361–73.

LANGER, R. H. M. (1966). 'Mineral nutrition of grasses and cereals', in *The Growth of Grasses and Cereals*, ed. Milthorpe and Ivins. Butterworths, London, 213–26.

LANGER, R. H. M. and OLUGBEMI, L. B. (1970). A study of New Zealand wheats. IV. Effects of extreme temperature at different stages of development. *N.Z. Jl agric. Res.* **13**, 878–86.

LANGER, R. H. M. and RYLE, G. J. A. (1958). Vegetative proliferations in herbage grasses. *J. Br. Grassld Soc.* **13**, 29–33.

LIEW, F. K. Y. (1968). Nitrogen supply at various stages of development in wheat. M.Agr.Sc. thesis, Lincoln College, New Zealand.

MARSHALL, C. and SAGAR, G. R. (1968). The interdependence of tillers in *Lolium multiflorum* Lam.—a quantitative assessment. *Ann. Bot.* (N.S.) **19**, 785–94.

MITCHELL, K. J. (1953). Influence of light and temperature on the growth of ryegrass (*Lolium* spp.). I. Pattern of vegetative development. *Physiologia Pl.* **6**, 21–46.

MOORE, C. W. E. (1964). 'Distribution of grasslands', in *Grasses and Grasslands*, ed. Barnard. Macmillan, London, 182–205.

MOORE, R. M. and BIDDISCOMBE (1964). 'The effects of grazing on grasslands', in *Grasses and Grasslands*, ed. Barnard. Macmillan, London, 211–35.

MORLEY, F. H. W. and SPEDDING, C. R. W. (1968). Agricultural systems and grazing experiments. *Herb. Abstr.* **38**, 279–87.

PALEG, L. G. (1960). Physiological effects of gibberellic acid. II. On starch hydrolizing enzymes of barley endosperm. *Pl. Physiol., Lancaster* **35**, 902–6.

RAWSON, H. M. and EVANS, L. T. (1970). The pattern of grain growth within the ear of wheat. *Aust. J. biol. Sci.* **23**, 753–64.

RAYMOND, W. F. (1969). The nutritive value of forage crops. *Adv. Agron.* **21**, 1–108.

ROBSON, M. J. and JEWISS, O. R. (1968). A comparison of British and North African varieties of tall fescue (*Festuca arundinacea*). II. Growth during winter and survival at low temperatures. *J. appl. Ecol.* **5**, 179–90.

RYLE, G. J. A. (1964). A comparison of leaf and tiller growth in seven perennial grasses as influenced by nitrogen and temperature. *J. Br. Grassld Soc.* **19**, 281–90.

RYLE, G. J. A. (1965). Effects of daylength and temperature on ear size in S.24 perennial ryegrass. *Ann. appl. Biol.* **55**, 107–14.

RYLE, G. J. A. (1967). Effects of shading on inflorescence size and development in temperate perennial grasses. *Ann. appl. Biol.* **59**, 297–308.

RYLE, G. J. A. and LANGER, R. H. M. (1963). Studies on the physiology of flowering of timothy (*Phleum pratense* L.). I. Influence of day length and temperature on initiation and differentiation of the inflorescence. *Ann. Bot.* (N.S.) **27**, 213–31.

SHARMAN, B. C. (1945). Leaf and bud initiation in the Gramineae. *Bot. Gaz.* **106**, 269–89.

SHARMAN, B. C. (1947). The biology and developmental morphology of the shoot apex in the Gramineae. *New Phytol.* **46**, 20–34.

SILSBURY, J. H. (1970). Leaf growth in pasture grasses. *Trop. Grasslds* **4**, 17–36.

SNAYDON, R. W. and BRADSHAW, A. D. (1961). Differential response to calcium within the species *Festuca ovina* L. *New Phytol.* **60**, 219–34.

STOY, V. (1966). *Funktion von Blatt, Halm und Ähre bei der Ertragsbildung vom Getreide.* Bericht über die Arbeitstagung 1966 der Arbeitsgemeinschaft der Saatzuchtleiter, Gumpenstein, 29–49.

THORNE, G. N. (1965). Photosynthesis of ears and flag leaves of wheat and barley. *Ann. Bot.* (N.S.) **29**, 317–29.

THORNE, G. N. (1966). 'Physiological aspects of grain yield in cereals', in *The Growth of Grasses and Cereals*, ed. Milthorpe and Ivins. Butterworths, London, 88–105.

WARDLAW, I. F. (1970). The early stages of grain development in wheat: response to light and temperature in a single variety. *Aust. J. biol. Sci.* **23**, 765–74.

WALKER, T. W. (1956). The nitrogen cycle in grassland soils. *J. Sci. Fd Agric.* **7**, 67–72.

WALLACE, H. A. and BRESSMAN, E. N. (1949). *Corn and Corn Breeding.* Wiley, New York.

WILLIAMS, R. D. (1962). On the physiological significance of seminal roots in perennial grasses. *Ann. Bot.* (N.S.) **26**, 126–36.

WILLIAMS, R. F. (1966). 'Development of the inflorescence in Gramineae', in *Growth of Grasses and Cereals*, ed. Milthorpe and Ivins. Butterworths, London, 74–87.